URBAN LANDSCAPES

The urban landscape is a valuable cultural asset. It is a major source of aesthetic stimulus, historical knowledge and a principal means of emotional and practical orientation. A great deal of public and private expenditure is concerned directly or indirectly with the creation and maintenance of urban landscapes; research in this field – from historical, geographical, architectural and planning perspectives – is flourishing. The time is ripe to integrate this knowledge in order to understand the cyclical processes of adaptation and renewal, different disciplinary and national approaches and the challenges of managing cities as a cultural asset whilst accommodating new forms and functions.

Urban Landscapes fulfils this need. Taking a multidisciplinary and a multinational approach, reflected in its authorship and the urban landscapes discussed, it addresses the principal academic and practical issues concerning the past, present and future of the built environment. The volume describes the historical development of urban landscapes and the roles of those who shape our cities. It offers an essential understanding of our present-day environment and argues for its enlightened management in the future.

J.W.R. Whitehand is Professor of Urban Geography at the University of Birmingham and **P.J. Larkham** is Lecturer in Planning at the University of Central England.

URBAN LANDSCAPES

International perspectives

Edited by

J. W. R. Whitehand
and
P. J. Larkham

London and New York

First published 1992
by Routledge
11 New Fetter Lane, London EC4P 4EE

Simultaneously published in the USA and Canada
by Routledge
a division of Routledge, Chapman and Hall, Inc.
29 West 35th Street, New York, NY 10001

© 1992 J. W. R. Whitehand and P. J. Larkham

Typeset in Scantext September by
Leaper & Gard Ltd, Bristol
Printed and bound in Great Britain by
Biddles Ltd, Guildford and King's Lynn

British Library Cataloguing in Publication Data

A catalogue record for this book is available from the British Library.

ISBN 0–415–07074–0

Library of Congress Cataloging-in-Publication Data
Urban landscapes: international perspectives/edited by J. W. R.
Whitehand and P. J. Larkham.
p. cm.—(Routledge geography and environment series)
Based on a conference held in Birmingham, England, in July 1990.
Includes bibliographical references and index.
ISBN 0–415–07074–0
1. City planning—History—20th century. 2. Architecture–
–Environmental aspects. I. Whitehand, J. W. R. II. Larkham, P. J.
(Peter J.), 1960– . III. Series.
NA9095.U74   1992   92-12622
711′.4′09—dc20   CIP

CONTENTS

Part II The Nature and Management of Urban Landscape Change

Conclusion

FIGURES

vii

FIGURES

TABLES

CONTRIBUTORS

Nigel J. Baker — Research Fellow, School of Geography, University of Birmingham, UK

Micha Bandini — Professor and Head of School of Architecture and Interior Design, North London Polytechnic, UK

David Friedman — Associate Professor of the History of Architecture, School of Architecture and Planning, Massachusetts Institute of Technology, USA

Deryck W. Holdsworth — Associate Professor, Department of Geography, Pennsylvania State University, USA

Andrew N. Jones — Chesterton Consulting Group, Milton Keynes; formerly of the School of Geography, University of Birmingham, UK

Paul L. Knox — Professor, College of Architecture and Urban Studies, Virginia Polytechnic Institute and State University, USA

Peter J. Larkham — Lecturer in Planning, School of Planning, University of Central England in Birmingham, UK

Anngret Simms — Senior Lecturer in Geography, Department of Geography, University College, Dublin, Eire

Terry R. Slater — Senior Lecturer in Geography, School of Geography, University of Birmingham, UK

Anne Vernez Moudon — Professor, Department of Urban Design and Planning, University of Washington, Seattle, USA

Joan Vilagrasa   Professor Titular, Head of Department of
                        Geography and History, University of
                        Lleida, Spain
J. W. R. Whitehand   Professor of Urban Geography, School of
                        Geography, University of Birmingham, UK

PREFACE

In July 1990 some forty-five scholars and practitioners from several countries met over two days in Birmingham to discuss recent developments in the study of urban landscapes. The fields of knowledge represented included geography, planning, history, architecture and design. The success of the conference reflected the considerable surge of interest in the physical appearance of towns and cities during the 1980s and it was evident to the organizers that many of the papers presented provided the basis for a volume of essays on the urban landscape that represented some of the more important work being undertaken on this topic in several different disciplines and countries. However, what is provided here is much more than the proceedings of the conference, for the papers presented on that occasion have been substantially revised and supplemented by other essays to give a more rounded, if not fully representative, treatment.

The organization of the conference and the preparation of this volume have been aided by a large number of individuals and organizations. Help received by individual authors is acknowledged at the ends of their chapters, but we should like to record here a number of general acknowledgements. The Nuffield Foundation, the British Council and West Midlands 90 contributed towards the costs of the travel and accommodation of a number of speakers at the conference whose contributions were subsequently rewritten to form individual chapters. The conference was also supported by two Study Groups of the Institute of British Geographers – the Historical Geography Research Group and the Urban Geography Study Group. We are grateful to the Chairman of the Urban Geography Study Group, Dr Michael Bradford, for suggesting that the volume form part of the Routledge *Urban Geography* series, and to fellow members of the Urban Morphology Research Group of the School of Geography at the University of

Birmingham, whose collective efforts ensured the success of the conference. Finally, we should like to record our indebtedness to Kevin Burkhill, Geoff Dowling and Simon Restorick for their assistance in the preparation of many of the illustrations for publication.

J. W. R. Whitehand, University of Birmingham
P. J. Larkham, University of Central England in Birmingham

INTRODUCTION

1

THE URBAN LANDSCAPE: ISSUES AND PERSPECTIVES

J. W. R. Whitehand and P. J. Larkham

This book brings together a wide range of perspectives on the urban landscape. In terms of its authorship and the urban landscapes discussed, it is designedly multidisciplinary and multinational. This wide scope reflects a growing awareness that progress in understanding and managing the built environment can be aided greatly by the integration of knowledge from different disciplines and different culture areas.

The study of the urban landscape, often known as 'urban morphology', has attracted the interest of scholars in a number of fields; most importantly in geography, but also in architecture, planning and, to a lesser extent, history. Within geography, urban morphology 'belongs as much to historical geography as to urban geography; a fact that reflects the longevity of the urban landscape that is the urban morphologist's object of study' (Whitehand 1987a: 250). In particular, its roots are in the morphogenetic research tradition of central Europe, dating back to the work of Schlüter. He postulated a morphology of the cultural landscape (*Kulturlandschaft*) as the counterpart in human geography of geomorphology within physical geography (Schlüter 1899). Within industrial countries, this made the urban landscape (*Stadtlandschaft*) a major research topic. Although Schlüter's direct influence extended little beyond the German-speaking countries, through the publications of the German *emigré* M. R. G. Conzen, who laid the foundations of urban morphogenetics in the English-speaking world, his indirect influence was much wider. In the German-speaking countries, the topic of urban form remains close to the mainstream of historical and urban geography, and there is less distinction there between the study of present-day towns and the study of their historical aspects than there is in the English-speaking world (Whitehand 1987a: 250).

1

Recently, the study of urban form has developed in several directions, among which the historical one is particularly strong. But interest in the historical development of urban landscapes has not been limited to scholars concerned primarily with the past. Much of the recent work by geographers and others interested in 'contextual' architecture and the planning, or management, of urban landscapes attaches considerable importance to the forms created by previous generations. Furthermore, urban morphologists are by no means limiting their attention to urban form narrowly conceived but are examining the individuals, organizations and processes shaping that form (Slater 1990a).

THE DECLINE AND RESURGENCE OF URBAN MORPHOLOGY

The history of urban morphology during the first half of the twentieth century, and its diverse research traditions, have been the subject of recent investigation (Whitehand 1981, 1987a, 1987b; Slater 1990b), much of it concerned with the urban morphogenetic tradition and the central role played in it by M. R. G. Conzen. The continued interest in the urban landscape among German-speaking geographers in the post-war period is evident in, for example, the major study of Vienna by Bobek and Lichtenberger (1966). Indigenous British urban morphologists have been less interested in conceptualizations of process than in description and classification, exemplified by Smailes's characterizations of present townscapes in broad terms, based on rapid reconnaissance surveys (Smailes 1955). In the United States, a significant school of cultural morphology developed in the late 1920s, largely independent of direct European influence. But this 'Berkeley School' was more productive in terms of research on rural landscapes than on urban landscapes (M. P. Conzen 1978: 130; Whitehand 1981: 12; 1987a: 255). All these research 'schools', shown schematically in figure 1.1, were still small-scale in their numbers of adherents and publications in the early 1960s.

In the later 1960s and early 1970s, research on urban form was less susceptible than many branches of geography to the 'quantitative revolution'. Nevertheless, this was a period when various quantitative methods were developed. Studies employing them were largely morphographic, describing physical forms rather than analysing their origins and development. They were largely ahistorical, even when they considered the survival and distribution of historical buildings (Davies 1968; Johnston 1969). Contemporary with this phase in the history of

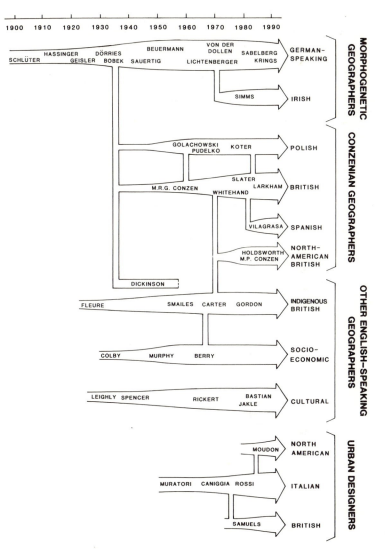

Figure 1.1 Research traditions within urban morphology: a schematic genealogy, showing a sample of authors.

geographical urban morphology were the development in the United States and widespread diffusion of concepts based on economics and the study of land-use patterns. The perspective of the urban geographers who adopted these concepts was 'morphological only in its concern with land-use patterns: town plan and building form were generally treated only as land-use containers, if considered at all' (Whitehand 1987a: 255). Additions to the number of researchers with a historical perspective on urban form were few. By 1970 urban morphology was characterized by Carter as a 'barren outpost of urban geography'. Fourteen years later, his view had apparently not changed greatly, for he regarded the subject as having been

> largely unaffected by those changing or shifting paradigms which supposedly have dominated geographical methodology. Quantitative analysis merely brushed ineffectually the periphery of morphological studies, while the present destruction of buildings is seen not in terms of its welfare consequences but rather in its impact on the cultural inheritance. More recent considerations of the structure of socio-political systems and their determinant organisation of space have again had little impact other than on the most general of scales.
>
> (Carter 1984: 145)

Somewhat earlier, however, M. P. Conzen had been able to detect a resurgence of research in urban morphology after the period of quiescence (M. P. Conzen 1978: 135). Publications dealing with the physical form of urban areas became more evident during the 1980s. Nevertheless, they formed only 12 per cent of geographical papers on the internal structure of cities in the middle of the decade (Whitehand 1986). In Britain, the major focus of geographical exploration of urban landscapes had become the Urban Morphology Research Group in the School of Geography at the University of Birmingham. A series of projects, broadly linked in methods and objectives, was undertaken (Larkham and Pompa 1988). The *Urban Morphology Newsletter*, edited by T. R. Slater, began regular publication in 1987. During the 1980s, growing contacts abroad had encouraged a revival, albeit limited, of urban morphological research in Poland, which had earlier received inspiration from the work of M. R. G. Conzen (Larkham 1987; Slater 1989b). In the late 1980s several researchers in northern Spain used approaches developed by M. R. G. Conzen and Whitehand; this research is sufficiently similar to that of the Birmingham Group for comparative projects to be undertaken (such as that by Vilagrasa

4

1990a). British urban designers and 'contextual' architects, occasionally spurred on by the interventions of HRH Prince Charles (Wales 1989; see also Jencks 1988), became increasingly aware of the significance of urban history and urban form in designing future urban landscapes. This resurgence in urban morphology, broadly defined, has occurred at much the same time as a renewed interest in the study of 'place' in geography (Johnston 1984, 1991). By 1990 it was felt that there was sufficient international interest in the urban landscape amongst academics and professionals in a range of disciplines for an International Conference on the Urban Landscape to be convened. The papers and types of research discussed on that occasion have shaped the contents of this volume.

THE IMPORTANCE OF M. R. G. CONZEN'S IDEAS

The most flourishing research tradition in geographical urban morphology, and the one with the widest distribution of adherents internationally, is that derived from the German morphogenetic school, introduced into Britain by M. R. G. Conzen (figure 1.1). This 'Conzenian' tradition deserves elaboration since, directly or indirectly, it permeates much of this book.

M. R. G. Conzen's upbringing and education in Berlin exposed him to a number of perspectives in the arts, humanities and natural sciences, and encouraged a remarkable breadth of vision. A geographer by initial training, he emigrated to Britain in 1933, on Hitler's accession to power, and became a professional town planner. During the Second World War he took up a lecturing appointment in the School of Geography at the University of Manchester under H. J. Fleure, moving to King's College, Newcastle (later the University of Newcastle upon Tyne) after the war, where he spent the remainder of his career (Whitehand 1987b; Slater 1990b).

Following the war, M. R. G. Conzen continued his research on urban morphogenesis. He produced a map of North-East England showing settlements classified by characteristics of form and period (M. R. G. Conzen 1949) and undertook detailed plot-by-plot surveys of a number of small British towns. He developed this type of work further, and applied his experience in planning, in his contribution to *A Survey of Whitby*, which was aimed at producing the basis for an integrated plan for the town (M. R. G. Conzen 1958). Evident in his contribution to this project were a concern for the conservation of period buildings and an interest in townscapes as composite historical artefacts (Larkham 1990: 352). His study of Alnwick (M. R. G. Conzen 1960), a comprehensive and detailed study of the town plan of a single town, was a further refinement of his survey technique. This monograph, which was innovative in conception and remarkable in its attention to the detail of the town plan, was regarded by the then editor of the Institute of British Geographers as 'undoubtedly one of the outstanding research publications of the Institute ... widely, and favourably, reviewed' (Steel 1984; see also Slater 1990b); it was reprinted in slightly modified form, with the addition of a glossary, in 1969 (M. R. G. Conzen 1969).

A significant factor in M. R. G. Conzen's contribution is his conceptualization of the way in which urban forms develop. His development of the concepts of the fringe belt and the burgage cycle and his tripartite division of the urban landscape into town plan, building forms and land use have been widely accepted as fundamental advances (Whitehand 1987a: 254). The fringe belt is a development of the *Stadtrandzone* identified by Louis (1936) in a study of Berlin. Fringe belts, simply described, are the physical manifestations in the landscape of periods of slow movement or even standstill in the outward extension of the built-up area; they tend to be occupied initially by land uses seeking large sites and having a low requirement for accessibility to the commercial core. The burgage cycle describes the progressive filling-in of plots with buildings, leading to a climax phase of maximum coverage and, ultimately, the clearance of plots preparatory to redevelopment.

M. R. G. Conzen returned to conservation as a theme in his paper on historical townscapes as a problem in applied geography (M. R. G. Conzen 1966), using as illustrations some of the small towns that he had surveyed in detail some years earlier. The concept of 'management' of the urban landscape was introduced, and the key attribute of an urban landscape in determining management priorities was identified as its historicity or historical expressiveness. The nature and intensity of the historicity of the urban landscape are expressed in practical terms by utilizing the division of the urban landscape into the three basic form complexes recognized by Conzen – town plan, building forms and land use. These are regarded as to some extent a hierarchy in which the building forms are contained within the plots or land-use units, which are in turn set in the framework of the town plan. These three form complexes, together with the site, combine at the most local level to produce the smallest, morphologically homogeneous areas that might be termed 'urban-landscape cells'. These cells are grouped into urban landscape units, which in turn combine at different levels of integration to form a hierarchy of intra-urban regions. The hierarchy of areal units is the geographical manifestation of the historical development of the urban landscape and encapsulates its historicity. It provides the reference point for all proposals for urban-landscape change (M. R. G. Conzen 1975). These ideas on conservation and historical townscapes are further discussed in Larkham (1990).

A number of current lines of research on urban form by geographers stem directly or indirectly from Conzen's ideas. Three of the most important are concerned with the nature and amounts of urban-landscape change, especially viewed over long time spans, the agents involved in the process of change, and the management of that change. In all cases there is a concern with features in the urban landscape that have been created by previous generations: the influence of the 'morphological frame' on subsequent developments is a recurrent theme.

The first of these lines of research is building directly upon the concern for history, through the analysis of historical, usually medieval, towns. A combination of historical documentation and plan analysis is leading to a more thorough understanding of the development of current urban landscapes (M. R. G. Conzen 1988). In particular, the practices of medieval town planning are being examined in detail by using, for example, the relative sizes and shapes of individual plots (or burgages) as clues to successive phases of planning, and by studying the differences between ideal and reality in the layout of towns (Slater 1987, 1988a, 1990d). Some of the towns that have been studied in this way are not commonly perceived as being of historical interest. Their medieval features may have been largely destroyed by industrial growth, as was the case with Wolverhampton and Doncaster (Slater 1986, 1989a).

In the second line of research, the study of urban landscapes has been linked more explicitly to the types of agents and the specific organizations and individuals responsible for their creation. Attention has largely been devoted to the period since the mid-nineteenth century, when sources permitting detailed building-by-building analyses became available in the form of building plans submitted to local authorities (Aspinall and Whitehand 1980). For the post-1947 period, similar data have been recovered from the records of local authority planning departments (Larkham 1988b). Using such data sources, reconstructions of urban development of unparalleled detail and completeness have been pieced together, sometimes for quite lengthy periods.

These types of detailed data have aided greatly a third strand of current geographical research in urban morphology. This is the concern for the planning, or management, of the urban landscape. Processes of decision-making are reconstructed, the agents (where surviving) are interviewed, and management procedures and policies are examined. This type of research has been successfully carried out on commercial cores and residential areas, with particular emphasis on conservation (Freeman 1988; Larkham 1988a, b; Whitehand 1990; Jones 1991). It merges with work in other disciplines, notably urban planning and design.

OTHER RESEARCH TRADITIONS

Largely separate from research stemming from the work of M. R. G. Conzen have been a number of other types of research on the urban landscape. These are rooted in several different disciplines (Whitehand 1992). Some of the publications in French and German (for example, Frantz 1987; Marois 1989) are general geographical studies of urban areas. They are part of a long-established tradition of geographical writing in which the urban landscape is only one of a range of phenomena to be considered. Also representative of a long-established tradition are continuing debates among English-speaking geographers, planners and economists on spatial variations in land values (Heikkila *et al.* 1989), particularly the rent-gap hypothesis (Clark 1988), which is frequently linked to the phenomenon of gentrification (Hamnett 1991).

However, the main concerns that are responsible for the growth of interest in the urban landscape are somewhat different. They include a dissatisfaction with what has been termed 'the passive view of human agency' by Ley (1988), and a sense that cities are undergoing fundamental change (Beauregard 1990; Knox 1991). Post-modernism, including its architectural manifestations, has been linked to a widespread economic restructuring. Doubts, extending well beyond urban morphology, have been cast on existing theories, but the types of features that have traditionally interested urban morphologists are to be found closer to the mainstreams of academic debate than for many years.

It is still far from clear how fruitful this part of the heightened interest in the urban landscape will prove to be. That a number of major, sometimes spectacular, urban-landscape features have been associated with post-modernism, for example high-tech corridors, festival settings and pedestrian shopping malls, is clear. And some of these have been the subject of individual examination (Hajdu 1988; Sawicki 1989; Hopkins 1990). But the need for a theory of late-twentieth-century urban landscapes distinct from earlier theories has yet to be demonstrated. Causal links between post-modern landscapes and economic restructuring have still to be convincingly shown. Although it is evident that, in the course of the 1970s, Western cities entered an architectural-style period distinct from that of the quarter century following the Second World War, it remains to be seen how far the connection between this fact and broadly contemporaneous social and economic changes involves relationships different from those associated with previous morphological periods.

Meanwhile, another type of investigation of urban form is attracting researchers of a quite different inclination. The use of computers to model and simulate urban physical structures is grounded in an intellectual milieu distinct from any so far mentioned. Most of this research has been undertaken by architects and geographers, the two groups working independently of one another. There is a need to relate this work to research of the types already described.

COMPARATIVE RESEARCH

During a video-recorded interview, M. R. G. Conzen was asked what he saw as the most pressing need in urban morphological research during the coming decade. The answer was

> to strengthen inter-disciplinary co-operation in relevant subjects and thereby to create the widest geographical basis of comparison; not only inter-disciplinary but international. ... In that way we can create an ultimately universal frame of reference for comparative study. And comparative study in subjects like geography, or history for that matter, is indispensable for the development and furtherance of conceptual thinking in these fields. You are the ideal researcher if in every individual case you can see both the individual as well as the general.
>
> (quoted in Slater 1988b)

The growth of the international body of researchers active in studies of urban form has not diminished the need for this broad, integrating view (Slater 1990c: 17). Recent conferences in Anglo-German urban historical geography, in which urban morphology played a significant part, revealed similarities between the concerns of British and German participants, but there are significant differences in both the questions asked and the research methods employed (see Denecke and Shaw 1988 for the publications stemming from two of these conferences). However, comparable research in different countries is achievable, as

9

has been shown in the exploration of concepts such as development cycles and the fringe belt (see von der Dollen 1978, 1990; Koter 1990; Vilagrasa 1990b), and projects based on data extracted from primary sources in more than one country are a practical proposition (Vilagrasa 1990a).

Whilst there are evident links between English-speaking and German-speaking urban morphologists, there are few between English-speaking and Italian-speaking architects. The work of Muratori and Caniggia has only recently achieved significant recognition among English-speaking scholars, notably through the publications of Samuels (1985, 1990). Muratori's theory and terminology have, however, received a thorough treatment by Malfroy (1986), who provides parallel texts in French and German. In view of the paucity of urban morphological research by French-speaking scholars, it is remarkable that the widest-ranging review of urban morphology, considering work in the English, German and Italian languages, appears in French (Choay and Merlin 1986). Even more remarkable, since one of the editors was a geographer, is the fact that no geographers were among the experts who took part in this survey, reflecting both the weak representation of urban morphology within French urban geography and the tenuousness of interdisciplinary communication.

INTERNATIONAL PERSPECTIVES ON THE URBAN LANDSCAPE

Historical urban landscapes

It is increasingly evident that both international and interdisciplinary co-operation are important in urban morphology. This is as true for studies of historical development as it is for studies of the nature and management of contemporary landscape change. The first of the two principal sections into which this book is divided deals with historical aspects of urban form, showing in a variety of ways how an appreciation of historical urban forms can be significant today. Not only do such forms shape the present urban landscape to a very considerable extent, but they provide cues for future urban planning and architecture.

In chapter 2, Simms, an historical geographer, begins the substantive part of the book with a development of earlier work on Romanized and non-Romanized Europe by herself and Ennen (Clarke and Simms 1985; Ennen 1985). She discusses the differences in urban origins and urban forms between these regions. These are demonstrated by the analysis of town plans. In some cases, such as Kells (Ireland), surviving buildings assist the recognition of plan units: in Kells the early monastic enclosure is readily visible in the current urban landscape. Simms provides a four-fold typology of towns based on major features of the early, or proto-town, phase of development. She emphasizes that a detailed study and interpretation of these town-plan features are of relevance to today's towns. She follows M. R. G. Conzen in valuing the historical elements of the urban landscape, and shows how these elements often reveal a continuity of function on the same site from early-medieval times to the present. Yet, since these are often central sites, they are particularly vulnerable to redevelopment, as in the case of Dublin's central area, and old road alignments are vulnerable to road-widening schemes. In showing how the past can influence the present, Simms sets the tone for the first section of the book.

Baker and Slater discuss an approach to the analysis of English medieval towns in chapter 3. The identification of distinct units within the town plan has increasingly been a part of the analysis of medieval towns during the 1980s, but the method for distinguishing these plan units is discussed at length for the first time in this chapter. The importance of the surviving elements of the urban landscape, in combination with cartographic, documentary and archaeological evidence, in analysing the evolution of the town plan is demonstrated using the example of Worcester, a cathedral town with a complex, multi-phase plan. It is emphasized that the precise dates of phases of town-plan evolution can seldom be discerned by this technique, but a relative chronology is often revealed. One of the important aspects of this growing body of research is the increased ability to draw parallels from analyses of other towns in the investigation and relative dating of particular plan units. This study reveals that plan units are not static, but change over time, and this is an important consideration when applying the results of the plan analysis of a medieval town to its future planning. The technique described by Baker and Slater not only provides a view of the development of the medieval town, but is a potential tool in the planning of the historic town in the late twentieth century.

Italian cities have long exerted a fascination for writers on urban form. In a valuable, early work, Rasmussen (1951) devotes a chapter to the replanning of Renaissance Rome. Siena is well known, and several examples of its buildings and plan elements are discussed in books on urban form and design: Logie (1954: 88–9), for example, illustrated the enclosure of the Piazza del Campo and its domination by the tower of the Palazzo Pubblico. More recently, Vance (1990: 139) has illustrated the 'frowningly powerful structure' of the Palazzo Ricardi Medici in Florence. But the processes that underlie the creation of forms are not explored in depth in any of these examples. In chapter 3, however, Friedman undertakes a detailed study of a crucial aspect of the development of late-medieval and Renaissance Italian towns, examining legislation and the reactions to its change over time, using documents of the period. His is the view of an architectural historian. He shows that the surviving buildings of the period, over which writers on urban form enthuse, were to a great extent faced with structures projecting forwards from the load-bearing masonry. These structures, often of wood, have vanished, leaving only minor scars on the buildings and transformed urban landscapes. Friedman discusses how this transformation occurred, at the end of the medieval period and the beginning of the Renaissance, at the same time that attitudes towards the front face of buildings – the façade – changed. By the end of the fifteenth century, the façade was an accepted aspect of urban culture and architecture. A new way of looking at, and designing, buildings had become accepted, which was to affect all buildings designed from that time to the present.

Holdsworth concludes the first part of the book with an examination of a quite different, but historically no less important, type of building transformation: the replacement of city centre buildings of a few storeys in height by skyscrapers, focusing on Manhattan during the short period 1893–1920. Despite including a world war, this period was one of enormous change to the urban landscape, with densities rapidly increasing as plots were amalgamated and as new technology led to higher buildings. Holdsworth's innovative use of a computer-based graphic design system to provide a time series of historical data on morphological change is a novel and effective way of portraying the changing urban landscape. This provides a basis for investigation of the corporate decision-makers and the social aspects of the growth of office working.

The nature and management of urban-landscape change

The second main section of this volume deals more explicitly with attitudes towards present urban landscapes. In particular, it considers how the urban landscapes of the late twentieth century arose and are managed. The examples described are drawn from several different countries, and range from suburbs to city centres.

In chapter 6, Bandini uses her familiarity with recent British and Italian architectural thought to chronicle the development of ideas in the post-war period in two countries whose traditions of thought on cities are very different. British thought is seen to be more empirically based than Italian: indeed, British architects took up theoretical constructs at a far slower rate, and later, than was the case elsewhere in Europe. Italian architects seem to have had a wider range of social and political concerns than the majority of British architects. The resulting fashions in thought that are described – including the 'Picturesque' debate, 'Team 10', and the New Town scheme – have close links with urban planning, and provide insight into the way in which the urban landscape is perceived by some of its influential makers.

Two chapters deal with different aspects of suburbia in the United States. Moudon, an urban designer, examines twentieth-century suburban development, particularly in the Seattle area, in chapter 7. Her examination of street networks, plots and building types is Conzenian in spirit, and enables her not only to suggest a classification of street networks and plot types, but also to identify morphological regions based on these characteristics. The older regions show the strong constraints of their original platting processes, which provided the initial lineaments within which developments proceeded. In contrast, regions created in the 1980s tend to have more open layouts, containing more footpaths and communal open space. This research suggests lessons for the design and management of urban landscapes. It shows the advantages of the narrow-plot house over the wide plot, the latter being an aberration stemming from easy access to artificially cheap land, and the restrictive effects of covenants that govern suburban landscapes and persist for many years after they have become outdated and irrelevant.

Knox develops the American theme by concentrating upon one particular residential type mentioned by Moudon. His chapter deals with private master-planned residential communities, a relatively recent innovation. These, termed 'artful fragments' by Boyer (1987), include the Planned Urban Development. The 'artfulness' in these suburban fragments comes in the variety and ostentation of their architecture and layout. Post-modernism and its associated pseudo-historical architectural styles are shown to good effect in the large and opulent houses. The estates in which they are located are expensively landscaped, have many communal facilities private to the estate, are often walled, and have private security services. This is a glimpse into the suburban landscape of the American élite. It is the most recent suburban manifestation of that 'restless urban landscape' about which Knox has written elsewhere (Knox 1991), and forms a significant development from his original paper to the conference.

13

Whitehand, Larkham and Jones continue the suburban theme with a consideration of post-war changes in established British suburban areas. Selecting six small areas from South-East England and the English Midlands for examination, they deal with a number of issues concerning the more intensive development of existing suburbs. Over one-half of the plots in some of the study areas have been the subject of attempts to undertake more intensive development. Densities have increased as plots have been subdivided for dwellings in the back, or even front, garden; culs-de-sac have been driven in to develop plot tails, and plots have been amalgamated for comprehensive redevelopment. What emerges from the majority of cases studied is the importance of conflict in the planning process, often resulting in outcomes in the urban landscape that are different from the original proposals by any of the parties involved. Conflict also leads to a protracted development process, with delays of years, or even decades, occurring between initial discussions and actual developments. It seems, from these case studies, that suburban landscapes, even those in Conservation Areas, are in general the subject of only very limited planning by the local authorities responsible for them.

Attention then turns from suburbs to city centres. Vilagrasa, in chapter 10, presents a comparison of changes to the commercial cores of Worcester (England) and Lleida (Spain) since 1945. This form of direct comparison is still a rarity in research on the urban landscape, and in this case clearly requires a grasp of the different national bureaucracies, both of them complex. Both commercial cores are located in cathedral cities of some historical and architectural merit, but they developed rather differently during the mid- and late twentieth century. Central Worcester shows similarities with a number of other English commercial cores, particularly in the introduction of modern architecture by non-local architects working for national retail chains, and in the supplanting of modern by post-modern in the 1980s. In central Lleida, the pattern of architectural styles and the agents involved are rather different, and Vilagrasa shows how the contrasting planning systems in the two countries have been responsible in part for these differences. In particular, the mechanistic application of set standards for building heights and volumes in Lleida led to a pattern of redevelopment very different from that in the core of Worcester, where a strong conservation movement had developed in the 1970s and 1980s. Vilagrasa concludes that central Worcester has clearly been subject to some degree of urban landscape management during recent decades, despite adverse reactions from local residents to particular developments. In contrast, Lleida has come only recently to a realization that its historical urban landscape requires active management.

Finally, chapter 11 identifies a number of central, current issues in urban morphology in the light of the preceding chapters. There is, of course, a strict limit to the range of issues that can be addressed in depth in a volume of this length and no attempt has been made to achieve comprehensiveness. Lacunae, such as Third World cities and industrial landscapes, are evident. However, what is offered is both a diverse set of perspectives on urban landscapes and a wide range of types of urban landscapes, ranging from commercial cores to suburbs, from the medieval to the post-modern, and from America to southern Europe. And, despite their diverse backgrounds, all the contributors share the conviction that the urban landscape is a vital aspect of academic research, with important implications for planning practice.

ACKNOWLEDGEMENTS

The authors would like to thank Professor M. R. G. Conzen for his helpful comments on a draft of this chapter. Part of this chapter is adapted from a paper by J. W. R. Whitehand in *Urban Studies* (1992).

REFERENCES

Aspinall, P. J. and Whitehand, J. W. R. (1980) 'Building plans: a major source for urban studies', *Area* 12, 3: 199–203.

Beauregard, R. A. (1990) 'Capital restructuring and the new built environment of global cities: New York and Los Angeles', *International Journal of Urban and Regional Research* 15, 1: 90–105.

Bobek, H. and Lichtenberger, E. (1966) *Wien: Bauliche und Entwicklung seit der Mitte des 19. Jahrhunderts*, Graz: Böklaus.

Boyer, C. (1987) 'The return of the aesthetic to city planning: future theory as a departure from the past', paper presented to the Conference on Planning Theory into the 1990s, Center for Urban Policy Research, Rutgers University.

Carter, H. (1970) 'A decision-making approach to town plan analysis: a case study of Llandudno', in Carter, H. and Davies, W. K. D. (eds) *Urban Essays: Studies in the Geography of Wales*, London: Longman.

Carter, H. (1984) review of Whitehand, J. W. R. (ed.) (1981) *The Urban Landscape: Historical Development and Management: Papers by M. R. G. Conzen*, Institute of British Geographers, Special Publication 13, London: Academic Press, in *Progress in Human Geography* 8, 1: 145–7.

Choay, F. and Merlin, P. (1986) *À Propos de la morphologie urbaine: tome 1 rapport de synthèse*, Noisy-le-Grand: University of Paris VIII.

Clark, E. (1988) 'The rent gap and transformation of the built environment:

case studies in Malmö 1860–1985', *Geografiska Annaler* 70B: 241–54.

Clarke, H. B. and Simms, A. (1985) 'Towards a comparative history of urban origins', in Clarke, H. B. and Simms, A. (eds) *The Comparative History of Urban Origins in Non-Roman Europe: Ireland, Wales, Denmark, Germany, Poland and Russia from the Ninth to the Thirteenth Century*, BAR International Series 255 (2 volumes), Oxford: British Archaeological Reports.

Conzen, M. P. (1978) 'Analytical approaches to the urban landscape', in Butzer, K. W. (ed.) *Dimensions of Human Geography: Essays in some Familiar and Neglected Themes*, Research Paper 186, Department of Geography, University of Chicago.

Conzen, M. R. G. (1949) 'Modern settlement', in Isaac, P. C. G. and Allan, R. E. A. (eds) *Scientific Survey of North-Eastern England*, Newcastle upon Tyne: British Association for the Advancement of Science.

Conzen, M. R. G. (1958) 'The growth and character of Whitby', in Daysh, G. H. J. (ed.) *A Survey of Whitby and the Surrounding Area*, Eton: Shakespeare Head Press.

Conzen, M. R. G. (1960) *Alnwick, Northumberland: a Study in Town-Plan Analysis*, Publication 27, London: Institute of British Geographers.

Conzen, M. R. G. (1966) 'Historical townscapes in Britain: a problem in applied geography', in House, J. W. (ed.) *Northern Geographical Essays in Honour of G. H. J. Daysh*, Newcastle upon Tyne: Oriel Press.

Conzen, M. R. G. (1969) *Alnwick, Northumberland: a Study in Town-Plan Analysis*, Publication 27, London: Institute of British Geographers (reprint, with minor revisions and the addition of a glossary, of a monograph first published in 1960).

Conzen, M. R. G. (1975) 'Geography and townscape conservation', in Uhlig, H. and Lienau, C. (eds) 'Anglo-German Symposium in Applied Geography, Giessen-Würzburg-München, 1973', *Giessener Geographische Schriften* 1975: 95–102.

Conzen, M. R. G. (1988) 'Morphogenesis, morphological regions and secular human agency in the historic townscape, as exemplified by Ludlow', in Denecke, D. and Shaw, G. (eds) *Urban Historical Geography: Recent Progress in Britain and Germany*, Cambridge: Cambridge University Press.

Davies, W. K. D. (1968) 'The morphology of central places: a case study', *Annals, Association of American Geographers* 58, 2: 91–110.

Denecke, D. and Shaw, G. (eds) (1988) *Urban Historical Geography: Recent Progress in Britain and Germany*, Cambridge: Cambridge University Press.

Dollen, B. von der (1978) 'Massnahmen zur Sanierung und Verschönerung der Altstadt Koblenz in der frühen Neuzeit', *Landeskundliche Vierteljahresblätter* 24: 3–15.

Dollen, B. von der (1990) 'An historico-geographical perspective on urban fringe-belt phenomena', in Slater, T. R. (ed.) *The Built Form of Western Cities*, Leicester: Leicester University Press.

Ennen, E. (1985) 'The early history of the European town: a retrospective view', in Clarke, H. B. and Simms, A. (eds) *The Comparative History of Urban Origins in non-Roman Europe: Ireland, Wales, Denmark, Germany, Poland and Russia from the Ninth to the Thirteenth Century*, BAR International Series 255 (2 volumes), Oxford: British Archaeological Reports.

Frantz, K. (1987) *Die Grossstadt Angloamericas im 18. und 19. Jahrhunderts,*

Erdkundliches Wissen 77, Stuttgart: Steiner.

Freeman, M. (1988) 'Developers, architects and building styles: post-war redevelopment in two town centres', *Transactions, Institute of British Geographers* NS 13, 2: 131–47.

Gebauer, M. and Samuels, I. (1981) *Urban Morphology: an Introduction*, Research Note 8, Joint Centre for Urban Design, Oxford Polytechnic.

Hajdu, J. G. (1988) 'Pedestrian malls in West Germany: perceptions of their role and stages in their development', *Journal of the American Planning Association* 54: 325–35.

Hamnett, C. (1991) 'The blind men and the elephant: the explanation of gentrification', *Transactions of the Institute of British Geographers* NS 16, 2: 173–89.

Heikkila, E., Gordon, P., Kim, J. I., Peiser, R. B., Richardson, H. W. and Dale-Johnson, D. (1989) 'What happened to the CBD-distance gradient? Land values in a policentric city', *Environment and Planning A* 21: 221–32.

Hopkins, J. S. P. (1990) 'West Edmonton Mall: landscape of myths and elsewhereness', *Canadian Geographer* 34, 1: 2–17.

Jencks, C. A. (1988) *The Prince, the Architects and the New Wave Monarchy*, London: Academy Editions.

Johnston, R. J. (1969) 'Towards an analytical study of the townscape: the residential building fabric', *Geografiska Annaler* 51B: 20–32.

Johnston, R. J. (1984) 'The world is our oyster', *Transactions of the Institute of British Geographers* NS 9, 4: 443–59.

Johnston, R. J. (1991) 'A place for everything and everything in its place', *Transactions of the Institute of British Geographers* NS 16, 2: 131–47.

Jones, A. N. (1991) 'The management of residential townscapes', unpublished PhD thesis, School of Geography, University of Birmingham.

Knox, P. L. (1991) 'The restless urban landscape: economic and socio-cultural change and the transformation of Washington DC', *Annals, Association of American Geographers* 81: 181–209.

Koter, M. (1990) 'The morphological evolution of a nineteenth-century city centre: Łódź, Poland, 1825–1973', in Slater, T. R. (ed.) *The Built Form of Western Cities*, Leicester: Leicester University Press.

Larkham, P. J. (1987) 'Urban morphology in Poland', *Urban Morphology Newsletter* 1: 3–4.

Larkham, P. J. (1988a) 'Agents and types of change in the conserved townscape', *Transactions of the Institute of British Geographers* NS 13, 2: 148–64.

Larkham, P. J. (1988b) 'Changing conservation areas in the English midlands: evidence from local planning records', *Urban Geography* 9, 5: 445–65.

Larkham, P. J. (1990) 'Conservation and the management of historical townscapes', in Slater, T. R. (ed.) *The Built Form of Western Cities*, Leicester: Leicester University Press.

Larkham, P. J. and Pompa, N. D. (1988) 'Research in urban morphology', *Planning History* 10, 2: 12–15.

Ley, D. (1988) 'From urban structure to urban landscape', *Urban Geography* 9, 1: 98–105.

Logie, G. (1954) *The Urban Scene*, London: Faber.

Louis, H. (1936) 'Die geographische Gliederung von Groß-Berlin', *Länderkund-*

liche Forschung Krebs-Festschrift: 146–71.

Lowndes, M. and Murray, K. (1988) 'Monumental dilemmas – and the development of rules of thumb for urban designers', *The Planner* 74, 3: 20–3.

Malfroy, S. (1986) 'Introduction à la terminologie de l'école muratorienne avec une référence particulière aux ouvrages méthodologique de Gianfranco Caniggia', in Malfroy, S. (ed.) *L'Approche morphologique de la ville et du territoire*, Zürich: Eidenössische Technische Hochschule.

Marois, C. (1989) 'Caractéristiques des changements du paysage urbain dans la ville de Montréal', *Annales de Géographie* 98: 385–402.

Rasmussen, S. E. (1951) *Towns and Buildings*, Liverpool: University Press of Liverpool.

Rock Townsend (1990) *Royal Leamington Spa: a Design Framework in an Historic Town*, Warwick: Warwick District Council, and London: English Heritage.

Samuels, I. (1985) *Urban Morphology in Design*, Research Note 14, Joint Centre for Urban Design, Oxford Polytechnic.

Samuels, I. (1990) 'Architectural practice and urban morphology', in Slater, T. R. (ed.) *The Built Form of Western Cities*, Leicester: Leicester University Press.

Sawicki, D. S. (1989) 'The festival marketplace as public policy: guidelines for future policy decisions', *Journal of the American Planning Association* 55: 348–61.

Schlüter, O. (1899) 'Bemerkungen zur Siedlungsgeographie', *Geographische Zeitschrift* 5: 65–84.

Slater, T. R. (1986) 'Wolverhampton: central place to medieval borough', in Slater, T. R. and Hooke, D. *Anglo-Saxon Wolverhampton and its Monastery*, Wolverhampton: Wolverhampton Metropolitan Borough Council.

Slater, T. R. (1987) 'Ideal and reality in English episcopal medieval town planning', *Transactions of the Institute of British Geographers* NS 12, 2: 191–203.

Slater, T. R. (1988a) 'English medieval town planning', in Denecke, D. and Shaw, G. (eds) *Urban Historical Geography: Recent Progress in Britain and Germany*, Cambridge: Cambridge University Press.

Slater, T. R. (1988b) 'Conversations with Con', *Area* 20, 2: 200–2.

Slater, T. R. (1989a) 'Doncaster's town plan: an analysis', in Buckland, P. C., Magilton, J. R. and Hayfield, C. (eds) *The Archaeology of Doncaster 2: the Medieval and Later Town*, BAR British Series 202(i), Oxford: British Archaeological Reports.

Slater, T. R. (1989b) 'Medieval and renaissance urban morphogenesis in eastern Poland', *Journal of Historical Geography* 15, 3: 239–59.

Slater, T. R. (ed.) (1990a) *The Built Form of Western Cities*, Leicester: Leicester University Press.

Slater, T. R. (1990b) 'Starting again: recollections of an urban morphologist', in Slater, T. R. (ed.) *The Built Form of Western Cities*, Leicester: Leicester University Press.

Slater, T. R. (1990c) 'Urban morphology in 1990: developments in international co-operation', in Slater, T. R. (ed.) *The Built Form of Western Cities*, Leicester: Leicester University Press.

Slater, T. R. (1990d) 'English medieval new towns with composite plans: evidence from the Midlands', in Slater, T. R. (ed.) *The Built Form of Western Cities*, Leicester: Leicester University Press.

Smailes, A. E. (1955) 'Some reflections on the geographical description and analysis of townscapes', *Transactions and Papers, Institute of British Geographers* 21: 99–115.

Steel, R. W. (1984) *The Institute of British Geographers: the First Fifty Years*, London: Institute of British Geographers.

Vance, J. E. Jr. (1990) *The Continuing City: Urban Morphology in Western Civilization*, Baltimore: Johns Hopkins University Press.

Vilagrasa, J. (1990a) *Centre Històric i Activitat Commercial: Worcester 1947–1988*, Espai/Temps 7, Lleida: Departament de Geografia i Història, Estudi General de Lleida.

Vilagrasa, J. (1990b) 'The fringe-belt concept in a Spanish context: the case of Lleida', in Slater, T. R. (ed.) *The Built Form of Western Cities*, Leicester: Leicester University Press.

Wales, HRH The Prince of (1989) *A Vision of Britain*, London: Doubleday.

Whitehand, J. W. R. (1981) 'Background to the urban morphogenetic tradition', in Whitehand, J. W. R. (ed.) *The Urban Landscape: Historical Development and Management: Papers by M. R. G. Conzen*, Institute of British Geographers, Special Publication 13, London: Academic Press.

Whitehand, J. W. R. (1986) 'Taking stock of urban geography', *Area* 18, 2: 147–51.

Whitehand, J. W. R. (1987a) 'Urban morphology', in Pacione, M. (ed.) *Historical Geography: Progress and Prospect*, London: Croom Helm.

Whitehand, J. W. R. (1987b) 'M. R. G. Conzen and the intellectual parentage of urban morphology', *Planning History Bulletin* 9, 2: 35–41.

Whitehand, J. W. R. (1990) 'Makers of the residential landscape: conflict and change in outer London', *Transactions of the Institute of British Geographers* NS 15, 1: 87–101.

Whitehand, J. W. R. (1992) 'Recent advances in urban morphology', *Urban Studies* 29: 617–34.

Part I

HISTORICAL URBAN LANDSCAPES

2

THE EARLY ORIGINS AND MORPHOLOGICAL INHERITANCE OF EUROPEAN TOWNS

Anngret Simms

INTRODUCTION

The early histories of towns have left physical signs in the urban land-scapes that we know today. These signs contribute significantly to our understanding of urban origins and provide links between present communities and their forebears. The imprint of the past is strongest in relation to the topographical organization of towns as expressed in the town plan. According to M. R. G. Conzen (1968: 116–17), the three fundamental form categories of the townscape are the town plan, the building fabric and the pattern of land and building utilization. Strictly speaking, the term 'town plan' means the cartographic representation of a town's physical layout to a predetermined scale. It consists of three elements: the association of streets in a street system, the individual plots and the ground-plan of the buildings. Of these distinct plan elements, the street system is the most persistent, providing the link with the early history of the town; it is, therefore, of the greatest value to the historical geographer and historian. However, archaeologists warn us to be cautious. Most present plans of medieval towns do not take us back to the original layout of a place at the time of foundation, but rather to a medieval phase, which was usually the result of various fundamental changes (Steuer 1988: 82).

The organization of space as expressed in the town plan has been a long-neglected dimension of urban life. Yet the topographical history of a town, its streets and buildings, can no more be separated from the history of urban society than one can separate the physical appearance of a human being from his or her personality (Reynolds 1977: 188). The international project for the publication of historic town atlases, focusing on plans of European towns at a unified scale of 1:2,500, will establish a good basis for future comparative research on European urban topography (Conzen 1976: 361–2).

URBAN ORIGINS

The most important time for the formation of towns in Europe was the medieval period, and investigation should, therefore, be focused here. In the eleventh century, economic growth and population increase led to the revival of towns, which were widely diffused throughout medieval Europe within a relatively short period. The roots of this extraordinary burst of economic and cultural energy lie in a variety of different historical conditions. While a large number of these towns were founded on completely new sites, others involved the superimposition of new structures on earlier beginnings. Of the greatest importance for any typology of early urban origins is the contrast between those parts of Europe that were once within the Roman Empire, and those that had stayed outside it. There is no doubt that those places where earlier Roman towns survived in some form had an easier start than others. Regarding the continuity of classical civilization, Ennen (1967: 174–82) has distinguished three broad zones of town life in medieval Europe:

1. the Mediterranean region, where classical urban traditions continued;
2. an intermediate zone stretching from western France, with an extension to England, along the Rhine and Danube regions, where Roman towns have left some traces; and
3. the area that was not directly influenced by Roman urban culture: the regions east of the Rhine, Scandinavia, Scotland and Ireland.

The question of continuity between the late Roman period and the early-medieval period touches the very roots of our European civilization. This is why there was so much passion in the debate initiated at the beginning of this century by the Austrian historian Dopsch, who promoted the theory that there was considerable continuity between the Roman and Germanic civilizations, in contrast to the concept that the barbarians owed little to the preceding classical period, and that Germanic civilization was, therefore, a completely new creation (Dopsch 1937). The difficult question is the nature of continuity between the Roman and medieval town. A simple answer is that it differed from place to place. The concept of continuity is meaningful only if we distinguish between different aspects: topographical, economic and cultural. Ennen (1985: 5) observed that no Roman town in the later German empire fully retained its antique form and importance. But almost all Roman *civitates* and many *castra* along the Rhine continued as settlements and contained remnants of the urbanized Gallo-Roman population. The most significant of those places fulfilled central-place functions as episcopal seats; for example, Cologne, Trier, Mainz, Speyer, Worms and Strasburg. This brings us to the essential point that the strongest link between the classical period and early-medieval times was the Christian Church, which saved the written character of classical culture and also continued the Roman tradition of an urban-based episcopacy. Accordingly, the episcopal seats along the Rhine, defended by the old Roman walls, became the model for the defended bishops' seats east of the Rhine. The idea of a town as a cult site and centre of ecclesiastical organization became one of the most powerful aspects of classical heritage, and spread all over Europe.

Cologne, with five hundred years of Roman history, is a good example of a town that constitutes a bridge between antiquity and Western civilization. When the Franks finally occupied the town in AD 460, they did not destroy it, but settled within its walls (Doppelfeld 1975). The regular Roman street pattern survived more or less intact to the present day (figure 2.1). In fact, the main axis of pedestrianization in the modern city follows the major Roman streets of the Cardo Maximus (Hohe Strasse) and Decumanus Maximus (Schilder Gasse). Cologne also illustrates the continuity of cult sites and the vital role of the Church in the transition period. The present-day cathedral stands on the site of the late Roman episcopal church which, in turn, was built upon the site of a Roman temple. The bishops took over the government of the town in the immediate post-Roman period. The continuity of site of secular administrative functions is equally impressive. The splendid Roman governor's palace lies underneath the medieval town hall and the modern city offices, so that the site from which Cologne is administered has not changed from Roman times to the twentieth century. The Roman city walls survived into the early-medieval period, and some sections are still extant (figure 2.1). A modern multi-lane motorway cuts through the historic centre of Roman and medieval remains in a north–south direction without any regard for the overall significance of that part of the city.

What about urban origins in non-Romanized Europe? The dynamic growth of economic life and population that led to the revival of towns in the once-Romanized regions of Europe led, in non-Romanized Europe, to the formation of a large number of new towns. More specifically, the increase in the power of lordship, growing territorial

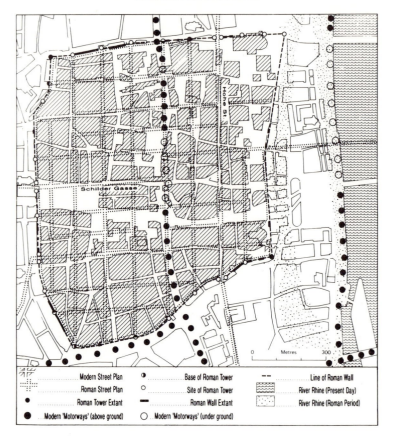

Figure 2.1 Cologne (Germany): the regularly laid-out Roman town superimposed on the modern street plan. Based on Römisch-Germanisches Zentralmuseum, Mainz 1990: 37, 2.

consolidation and the emergence of groups of craftsmen and traders under the protection of either secular or ecclesiastical lords contributed to the formation of towns (figure 2.2). Production and exchange of goods took place within the *suburbia* (small settlements of merchants and craftsmen), which grew up outside the early-medieval strongholds. In the twelfth and thirteenth centuries, the fully fledged medieval town with an urban constitution was superimposed upon these incipient forms of towns, which can be referred to as proto-towns.

A typology of these proto-towns emphasizes the primary formative

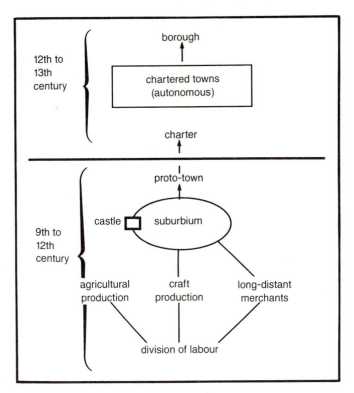

Figure 2.2 Interplay of processes that contributed to the formation of towns.
Based on Clarke and Simms 1985.

process in each case. Four categories can be distinguished (Clarke and Simms 1985: 678–89) (figure 2.3):

1. stronghold settlements (e.g. Dublin, Hamburg, Krakow),
2. coastal trading settlements (e.g. Haithabu, Birka, Kaupang),
3. market settlements (e.g. Oslo, Magdeburg, Frankfurt/Oder), and
4. cult settlements (e.g. Kells, Fritzlar, Odense).

These categories should not blind us to the fact that many of the most successful of these towns owed their development to their location on an important long-distance trade route, as well as to the existence of a stronghold providing protection for the growing merchant community.

It is noticeable that coastal trading places without the protection of lordship had little chance of survival: once-flourishing settlements such

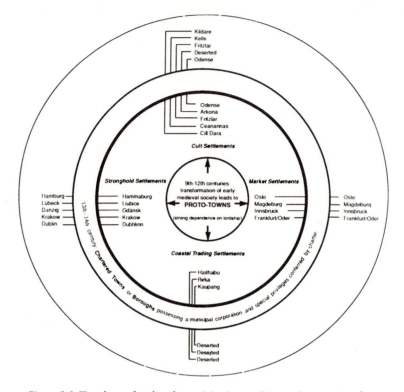

Figure 2.3 Typology of early urban origins in non-Roman Europe. Based on
Clarke and Simms 1985.

as Haithabu, Birka and Kaupang in Scandinavia were deserted.
Stronghold settlements in coastal locations (such as Dublin and
Hamburg) and on long-distance routes (Krakow), and market settlements, which were most frequently also strongholds, had the best
chance of success.

These proto-towns lacked a specific town law and an urban constitution, and they therefore lacked the autonomous administration of the
market and the town court. These attributes were to come with the
institution of the town charter, which granted burgesses the privilege of
self-government. The concept of the chartered town was borrowed from
formerly Romanized Europe. The oldest-known surviving charter north
of the Alps is that of Huy, in present-day Belgium, dated 1066. The

constitutional ideas of the town charter were reflected in the town plan, the market area for the activities of the burgesses being the focus of the town. The town walls dramatically represented the need for communal protection. In regions where the town walls enclosed the total urban community, they also symbolized the fact that there was a legal difference between townspeople and people from the country. In those instances, the distinction between proto-towns and chartered towns was, in fact, not only conceptual but also topographical.

The granting of town charters across Europe was closely linked to the various medieval colonization movements, which helped to spread cultural innovations (figure 2.4). At the western end of Europe, Ireland experienced urbanization under the influence of colonization, repeating the historical conditions experienced by people at the eastern periphery of the medieval core area. From the late twelfth century onwards, the Anglo-Normans colonized large parts of Ireland, and simultaneously the Germans colonized the lands of the West Slavic peoples east of the River Elbe. The foundation of chartered towns was a powerful tool in the hands of the colonizers, and helped to bring both regions into line with developments in the feudal core areas of medieval Europe.

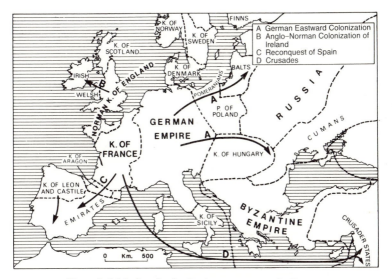

Figure 2.4 Medieval colonization movements in Europe.
From Simms 1988: 23.

URBAN ORIGINS AND TOWN LAYOUT

The question that specifically concerns us here is how these two major phases of urban development, the proto-urban phase and the autonomous town, influenced the layout of towns. This can be illustrated with case studies from Ireland and Central Europe.

Kells, a small market town in Ireland, is a good example of a cult centre that functioned as a proto-town and developed into a fully-fledged medieval town (Simms with Simms 1990). Kells was an important early Christian Columban monastery, which flourished between the ninth and eleventh centuries. Charters written in the Irish language on blank pages in the famous illustrated gospel book of Kells show the existence of a large secularized monastery with 'clergy and laity, freemen and strangers'. The charters refer to a school, a guest-house, craftsmen, agricultural tenants and, most importantly, the *margad*, a Scandinavian loan-word meaning market. This demonstrates that Kells functioned as a proto-town before the Anglo-Norman conquest. Traces of this incipient town are preserved in the alignment of the modern street pattern, which follows the curve of the outer- and inner-monastic enclosure containing the medieval church and church-yard (figure 2.5). The more secular part of the monastery was developed south-eastward, where the market was held just outside the eastern gate. Located between the two monastic enclosures stands Columb's House, a stone building dating from *c.* AD 1000. Just within the church-yard, the most imposing reminder of the monastic town is the tall, slim, round tower dating from *c.* AD 1100 (figure 2.6). Kells is typical of other monastic sites which appear to have been designed in conformity with a planned arrangement, in which the round tower usually stands to the west of the church. The entrance to the enclosure was generally located to the east and was marked by a special cross, a boundary cross, around which market functions developed. When the Anglo-Normans came, Kells was granted a charter based on the law of Breteuil, a small town in Normandy, and was extended along the perimeter of the monastic site. Obviously, no town-planning activity should interfere with the alignment of the former circular enclosures, which recall the structure of the earlier monastic town. Fortunately, the ambitious plans of an eighteenth-century landlord, that would have interfered with the alignment of the former enclosures, failed owing to a lack of funds.

Dublin is an excellent example of a medieval chartered town that took its origin from a Viking stronghold settlement with coastal trading functions (figure 2.7). The walled town is situated on high ground south

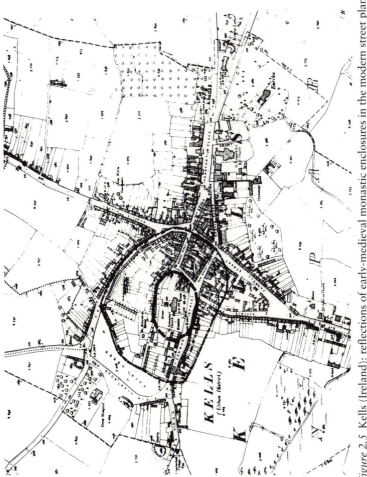

Figure 2.5 Kells (Ireland): reflections of early-medieval monastic enclosures in the modern street plan. Based on Ordnance Survey 1:2,500 sheet 17, edition of 1908–11.

Figure 2.6 Kells (Ireland): aerial view showing round tower, church, cross and circular alignment of road. Photograph by Michael Heriby.

of the River Liffey. The eastern section, around the castle, is the oldest. It was here that the Vikings erected their stronghold in the tenth century. From the early eleventh century the Hiberno-Norse population extended their settlement westward along the road on the crest of the hill, and later enclosed their settlement with a stone wall. After the coming of the Anglo-Normans, the English monarch Henry II granted Dublin to his men of Bristol by a charter issued in 1171–2. The walled town was extended into the river flood-plain to gain a greater depth of water along the quays for larger ships. During the fourteenth century, at a time of crisis, the town wall was extended along the quays.

Viking Dublin has considerable locational and topographical similarities to coastal stronghold settlements such as Lubeck and Szczecin (Stettin). All three towns are based on proto-urban settlements, which were succeeded by chartered medieval towns. All three are located at the confluence of a smaller river with a larger one, and on a hill overlooking a river. But this is where the similarities end. While Lubeck has declared its medieval core to be an architectural monument, Dublin Corporation has diminished its medieval town by building massive, high-rise office blocks next to the medieval cathedral, in the heart of the old Viking

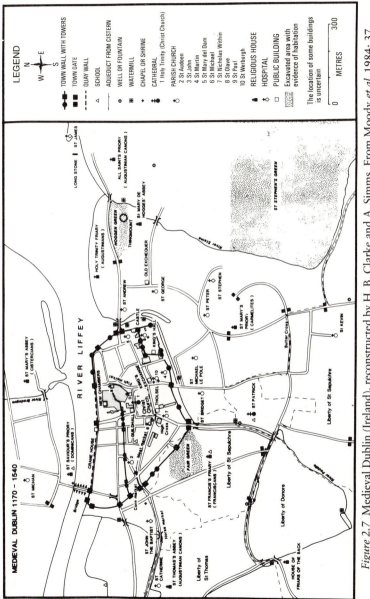

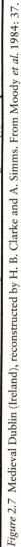

Figure 2.7 Medieval Dublin (Ireland), reconstructed by H. B. Clarke and A. Simms. From Moody *et al.* 1984: 37.

town (figure 2.8). In the process of building these office blocks, the earliest-known enclosures of Dublin, the Viking embankments dating from the tenth century, were levelled. These would have been the best testimony to the early origins of the town as a stronghold settlement. Dublin Corporation also carried out insensitive road-widening schemes that cut right through the medieval town from north to south (figure 2.9). Prolonged planning blight in the area has allowed speculative developers to assemble large sites, many of which have recently been developed with the assistance of tax-relief schemes. For example, the original plot pattern south of High Street is completely obliterated and has been replaced by a single large office block redevelopment.

A last Irish example is Drogheda, on the east coast of Ireland. Drogheda is a typical colonial town, founded by the Anglo-Normans, having no earlier urban roots. The town plan reflects its feudal history (figure 2.10). The Anglo-Norman lord here was Hugh de Lacy, who built his earthen stronghold (motte) here before 1186. The earliest surviving charter, based on the law of Breteuil, dates from 1194. This charter confirms that each burgess may hold a plot in which the front is 50 feet wide. Within two generations, between *c.* 1186 and *c.* 1230, the town was regularly laid out with intersecting streets so that the plan

Figure 2.8 Wood Quay Civic Offices, Dublin.

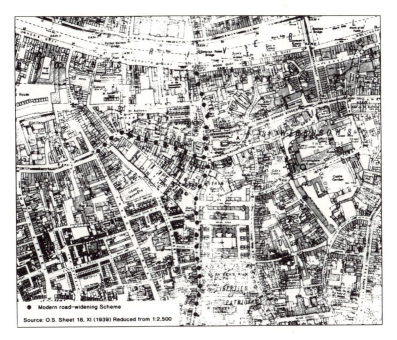

Figure 2.9 Modern road-widening schemes in the medieval core of Dublin.
Based on Ordnance Survey 1:2,500 sheet 18, edition of 1939.

approximated to a grid. Bradley (1978) confirms that much of the
present plot pattern within the walled town is a medieval survival.
Together with the street pattern, this provides an example of a town's
past that has survived at ground level, rather than above or below it.
The main north–south street served as the market place. A major road
bypass has mercifully left the original town plan of Drogheda intact.

Chronologically, the Irish monastic proto-towns and Viking towns
from the tenth century should be seen as being contemporary with the
early trading places of the Western Slavs along the Baltic coast and
the major rivers, for example the Oder and the Vistula. Similarly, the
formation of chartered towns in the context of medieval colonization
from the twelfth century onwards occurred simultaneously on the
western and eastern periphery of the medieval core area of Europe.

Opole (Oppelin), on the River Oder in Poland, is a good example of a
Slavic stronghold and civitas from the ninth to thirteenth centuries, with
an enclosed *suburbium* of craftsmen and merchants, which was later
replaced by a chartered town deliberately founded by a Slavic duke

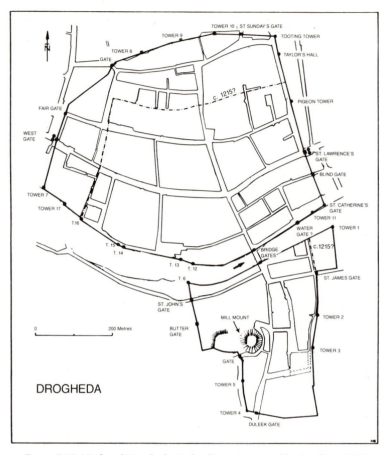

Figure 2.10 Medieval Drogheda (Ireland), reconstructed by Bradley (1978: 119).

(Kuhn 1979). Archaeological excavations in the Slavic stronghold have produced fragments of silk, glass and ivory. The settlement contained some one hundred wooden houses. In the early thirteenth century, the Polish Duke Kasimir invited German settlers to the town and granted them a charter according to the town law of Magdeburg, with market rights and liberties. A new town was then built, which attracted settlers away from the old Slavic stronghold. The new town began with a settlement around the Church of the Holy Cross, and, a few decades later, a regularly laid-out settlement with a grid pattern was created around a central market place. The house plots ran parallel to each other and had

36

a width of 8 to 9 metres, running from one street to the next. Within the new town, there was considerably more space available than there had been in the old Slavic stronghold. The centrally placed market square with its town hall is a good indication that the craftsmen and merchants were the driving force in this town. They traded cloth and ale, which they produced within their houses, and salt.

Both Drogheda and Opole show the imprint of a feudal lord residing in a castle across a river from a walled and regularly laid-out thirteenth-century town, with an emphasis on the market area. Both towns were granted charters modelled on towns within the old feudal core area of Europe.

Rostock (figure 2.11), on the Baltic in the former East Germany, is a fine example of how a colonial town can result from different growth stages, rather than being the result of one single act or phase of planning (Clarke and Simms 1985: 693–4). In the middle of the twelfth century, Rostock was a Slavic stronghold settlement on the eastern bank of the River Warnow, opposite the site of the modern town. In 1161 the stronghold was destroyed during the Wendish crusades, and the Slavic prince, Pribislav, submitted to Henry the Lion and became his feudal sub-tenant. The immigration of German settlers began. As the Slavic *wik* (trading post) became too small, the settlers were granted land across the river, on the steep western bank of the Warnow. Here, the Church of St Peter was built in association with the present Alter Markt. In 1218, the inhabitants of this market area were granted a town charter according to the Law of Lübeck by Prince Henry Borwin I. By 1232 a new quarter around the Church of St Mary and the Neuer Markt had been established, and a third nucleus was formed by 1252 around the Church of St Jacob and the Hopfen Markt. Within four decades, three independent quarters had developed, each with its own market and church. In 1265, a new charter was issued, which extended town law over all three existing settlement nuclei, and the street system was changed into what now appears to be a quite regular layout, in order that the different parts of the town could be integrated.

Similar developments occurred in other coastal towns on the Baltic that received the Law of Lübeck, for example in Gdansk (Danzig) and Kaliningrad (Königsberg). The major features of Rostock's early town plan, the churches at the core of new urban quarters on the one hand, and the market in the centre of the town on the other hand, have had an impact upon the urban landscape lasting well beyond their medieval origin. The churches, albeit reconstructed after the damage of the last war, still dominate the townscape. The buildings around the market

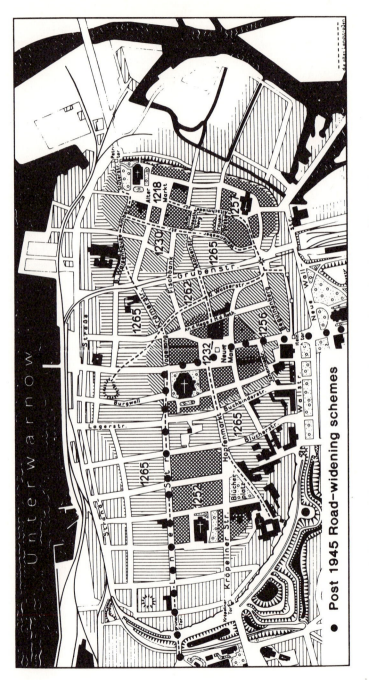

Figure 2.11 Post-war road-widening schemes in the medieval core of Rostock (Germany).

place have been rebuilt, but the unity of the square has suffered from having a major modern road run straight through it (figure 2.11). The original architectural model for the town hall and market place was Lübeck. The oldest, late-medieval houses in Rostock with stepped gables, and later ones with Flemish gables, were also modelled on those in Lübeck. We see here a very far-reaching cultural diffusion process, which has stimulated modern town planners to design a new Hanseatic style of house, reminiscent of its predecessors. The transformation of the old hop-market by a large fountain is a good example of an attempt by modern planners to recreate public space, a central concept in a medieval town.

URBAN ORIGINS AND MODERN URBAN LANDSCAPES

The historic cores of our cities are an irreplaceable cultural resource (Conzen 1975). A significant element in this is the town plan, which in itself illustrates the history of the physical growth of a town. It is necessary that the primary dynamic in the evolution of a town is well understood, so that the physical expression of early urban growth is fully recognized and can be integrated into modern town plans, in order to build bridges to the past and to strengthen a town's identity. As long as planners are unaware of town-plan analysis, they will fail to appreciate fully the importance of old street systems and they will, therefore, continue to follow a narrow townscape approach that leads to an emphasis solely on façades and the protection of designated buildings, instead of the recognition of characteristic quarters in the larger town plan, each with its specific infrastructure, of which the old road system would be a defining element (Slater 1984: 333). Ironically, the town where this definition of distinct historical quarters appears to have been most successfully achieved by modern townscape management is Boston, Massachusetts, where a very obvious alliance has been struck between historical quarters and the middle classes (Whitehall 1990). A major drawback in what is otherwise a success story is the separation of the oldest quarter from the rest of the city by a multi-lane urban motorway, although there are plans to put this major roadway underground.

It should be possible to conceptualize the different morphological expressions of early urban growth better than we have been able to do thus far, and thereby to make a contribution to an historico-geographical theory of urban form, the development of which Whitehand (1977) identified as one of the most urgent tasks of modern morphological studies. This aim is probably best achieved by interdisciplinary co-operation between historical geographers, historians and architects. Such co-operation should be a step towards establishing an accord between historical knowledge and contemporary decisions affecting the management of historic townscapes.

The plea to allow the centres of historic towns to have an urban form that does justice to their own origins is not an attempt to use the medieval town as an instrument of conservative social criticism. As living entities, and subject to cultural, economic and social evolution, historic towns and districts must inevitably change, as they have done in the past. However, historic towns should not solely meet the management demands of contemporary life, but should also ensure the preservation of cultural, human and architectural values.

The need to take root in a town by identifying with its history has a particular urgency in the post-modern world. The almost exclusive concentration on the diversity of modern cultural expressions and the rejection of any official culture as inherently élitist and oppressive have put into question the possibility of a coherent civic culture, which could be shared (Cosgrove 1990: 563). In our roles as consumers, we participate in aspects of urban culture which may be inaccessible to other groups of people although they live in one and the same town. In contrast, the central images of a medieval town, whether we think of city walls or churches, would have been accessible and appreciated by all citizens. Modern society has to cope with a multiplication of meaning; but, without shared meaning, a public place loses its significance and vandalism is a terrible consequence. A town without a historic landmark of any kind, even if it is relatively recent, is deeply frightening: it is like meeting a person without a memory. We instinctively expect the history of a town to be the biography of its people, a collective personality, which has expressed itself over time in the built environment. The town plan reflects the living and working conditions of medieval townspeople more broadly than individual great buildings, public or private, that may have been selected for conservation. If there are clear images of the successive lives of the town from its early beginnings to the present in the physical fabric of the town, then they can maintain or revive some consciousness of citizenship. There is most certainly a public, and indeed a civic, dimension in our endeavours of establishing a typology of early urban forms, which would contribute towards a sensitive and culturally aware management of historic townscapes.

ACKNOWLEDGEMENT

Figure 2.7 was previously published in Moody, T. W., Martin, F. X. and Byrne, F. J. (eds) (1984) *A New History of Ireland, ix: maps, genealogies, lists*, Oxford: Oxford University Press.

REFERENCES

Böhme, H. (1987) 'Stadtebau als konservative Gesellschaftskritik', *Die Alte Stadt* 14, 1: 1–27.

Bradley, J. (1978) 'The topography and layout of medieval Drogheda', *County Louth Archaeological and Historical Journal* 19, 2: 98–126.

Clarke, H. B. and Simms, A. (1985) 'Towards a comparative history of urban origins', in Clarke, H. B. and Simms, A. (eds) *The Comparative History of Urban Origins in Non-Roman Europe: Ireland, Wales, Denmark, Germany, Poland and Russia from the Ninth to the Thirteenth Century*, BAR International Series 255 (2 volumes), Oxford: British Archaeological Reports.

Conzen, M. R. G. (1968) 'The use of town plans in the study of urban history', in Dyos, H. J. (ed.) *The Study of Urban History*, London: Edward Arnold.

Conzen, M. R. G. (1975) 'Geography and townscape conservation', in Uhlig, H. and Lienau, C. (eds) 'Anglo-German symposium in applied geography, Giessen-Würzburg-München, 1973', *Giessener Geographische Schriften* 1975: 95–102.

Conzen, M. R. G. (1976) 'A note on the Historic Towns Atlas', *Journal of Historical Geography* 2, 4: 361–2.

Cooke, P. (1990) 'Modern urban theory in question', *Transactions of the Institute of British Geographers* NS 15, 3: 331–43.

Cosgrove, D. (1990) '. . . Then we take Berlin: cultural geography 1989–90', *Progress in Human Geography* 14, 4: 560–8.

Denecke, D. (1989) 'Stadtgeographie als geographische Gesamtdarstellung und komplexe Analyse der Stadt', *Die Alte Stadt* 16: 3–23.

Doppelfeld, O. (1975) 'Köln von der Spätantike bis zur Karolingerzeit', in Jankuhn, H., Schlesinger, W. and Steuer, H. (eds) *Vor- und Frühformen der europäischen Stadt im Mittelalter*, Göttingen: Vandenhoeck and Ruprecht.

Dopsch, A. (1937) *The Economic and Social Foundations of European Civilization*, London: Kegan Paul.

Ennen, E. (1967) 'The different types of formation of European towns', in Thrupp, S. L. (ed.) *Early Medieval Society*, New York: Appleton-Century-Crofts.

Ennen, E. (1985) 'The early history of the European town: a retrospective view', in Clarke, H. B. and Simms, A. (eds) *The Comparative History of Urban Origins in Non-Roman Europe: Ireland, Wales, Denmark, Germany, Poland and Russia from the Ninth to the Thirteenth Century*, BAR International Series 255, volume 1, Oxford: British Archaeological Reports.

Gruber, K. (1977) *Die Gestalt der Deutschen Stadt*, Munchen: Verlag Georg Callwey (first edition 1937).

ICOMOS, Eger International Committee on Historic Towns (1986) *International Charter for the Protection of Historic Towns*, Washington, DC:

International Council on Monuments and Sites.

Kuhn, W. (1979) 'Oppelin', in Stoob, H. (ed.) *Deutscher Städteatlas*, volume 2, Münster: Institut für vergleichende Städtegeschichte.

Moody, T. W., Martin, F. X. and Byrne, F. J. (eds) (1984) *A New History of Ireland*, volume 9, Oxford: Clarendon Press.

Reynolds, S. (1977) *An Introduction to the History of English Medieval Towns*, Oxford: Clarendon Press.

Römisch-Germanisches Zentralmuseum, Mainz (1990) *Führer zu vor- und frühgeschichtlichen Denkmälern*, Mainz: Philipp Zabern.

Secchi, B. (1976) 'The new quality of the question of historic centres', *Lotus International* (December) (unpaginated).

Simms, A. (1988) 'Core and periphery in medieval Europe: the Irish experience in a wider context', in Smith, W. J. and Whelan, K. (eds) *Common Ground*, Cork: Cork University Press.

Simms, A. with Simms, K. (1990) 'Kells' in Andrews, J. and Simms, A. (eds) *Irish Historic Towns Atlas*, volume 4, Dublin: Royal Irish Academy.

Slater, T. R. (1984) 'Preservation, conservation and planning in historic towns', *Geographical Journal* 150, 3: 322–34.

Steuer, H. (1988) 'Urban archaeology in Germany and the study of topographic, functional and social structures', in Denecke, D. and Shaw, G. (eds) *Urban Historical Geography: Recent Progress in Britain and Germany*, Cambridge: Cambridge University Press.

Whitehall, W. M. (1990) *Boston: a topographical history*, 3rd edition, Cambridge, MA: The Belknap Press of Harvard University Press.

Whitehand, J. W. R. (1977) 'The basis for an historico-geographical theory of urban form', *Transactions of the Institute of British Geographers* NS 2, 3: 400–16.

3

MORPHOLOGICAL REGIONS IN ENGLISH MEDIEVAL TOWNS

N. J. Baker and T. R. Slater

The analysis of the physical characteristics of English towns through devising morphologically based regions has its origins in the research of M. R. G. Conzen (1960, 1962). That research, in turn, is a development of the work of central European researchers in the period before 1945 (for example, Geisler 1924; Strahm 1935). Conzen's papers on Alnwick, Northumberland and the city of Newcastle upon Tyne, whilst conceptually rich and, in many ways, extremely detailed, concentrated on the morphological evolution of the town plan in the early-modern, industrial and twentieth-century periods. They do not provide a detailed exposition on how to devise morphological regions, nor do they concentrate unduly upon the medieval development of the core of the towns. Subsequent work by Conzen, most notably upon the town of Ludlow, Shropshire (Conzen 1968, 1988), has expanded upon his earlier work in the context of a place with much more distinctive plan divisions deriving from its medieval phase of development but, again, techniques for deriving morphological regions are not dealt with as such. He does, however, expand upon the concept of the morphological region in towns. This, he says, is a combination of the regionalized pattern of the three systematic form complexes of town plan, building fabric and land use. Of these, the town plan (street and plot pattern) is the most resistant to change and the pattern of land utilization the most liable to change, particularly over the past 150 years. The building fabric is the most visually obvious carrier of historical information in the townscape and, because it represents a major fixed capital investment on the part of individuals, has also been reasonably resistant to major change until comparatively recently, though not to the extent of the town plan (Conzen 1988: 255–9).

Research by one of the present authors (Slater 1981, 1985, 1988) has utilized this analytical framework to investigate a variety of smaller medieval towns, mostly in the English Midlands, concentrating particularly upon the plan characteristics of plot series, the evidence for town planning in the medieval period, and medieval development processes. Apart from this, there has been little research upon the plan development of medieval towns based upon cartographic sources. Historical and archaeological sources have been used to reconstruct medieval townscapes, but rarely for the whole of a town and rarely upon an accurate cartographic base. Keene's (1985) survey of medieval Winchester is the most substantive of such studies, but this makes no attempt to look in detail at the morphological development of the plan of the city through the period; rather, the focus is on social and economic development. The three volumes of the *British Historic Towns Atlas* (edited by Lobel 1968, 1975, 1990) have an overtly topographic purpose, namely to reconstruct the plans of towns for the period just before the phase of industrial growth at a standard scale of 1:5000. However, unlike some of their European counterparts, the British volumes have no analysis of the development of the plans as reconstructed. Questions have also arisen as to the accuracy of the reconstructions that have been published (Slater and Lilley, forthcoming). The few published accounts of the landscape of medieval towns which have attempted to use the regionalization procedures developed by Conzen, other than those by Slater already referred to, include a study of Thame, Oxfordshire, by Bond (1990); an analysis of Perth, Scotland by Spearman (1988); the investigation of some Anglo-Saxon towns in southern England by Haslam (1985); an analysis of the pre-industrial development of Walsall (Baker 1989); and a number of Irish town studies by Bradley (1985, 1990) and Simms (1990).

PLAN UNITS

The primary division of Conzen's analytic framework for the town plan was between the medieval core region called 'the old town' (after the German *Altstadt*), or the 'kernel', and accretions from the sixteenth century onwards. Here the key concept was that of the fringe belt, an area whose plan characteristics derive from land uses that seek peripheral locations in the town during periods of standstill or only slowly advancing growth. The secondary subdivisions of the plan of the old-town core were conceptualized by Conzen as being in two orders of magnitude. Those of the lowest order represent parcels of old-town plots which have a high degree of internal homogeneity in their plan characteristics, most notably in lengths and breadths of plots. Those of the second order of magnitude represent amalgamations of these plot parcels into more substantive plan divisions, such as the combination of the three slightly different plot series that surround the market place at Alnwick, or those that make up each of the extra-mural suburbs of that town. It is these second-order plan units that form the basis of Conzen's analysis of Ludlow and his hypotheses about the chronological development of the morphological regions discerned there. Subsequent, more detailed, work on the plan of Ludlow, the historical documentation of the town, and some archaeological investigation have suggested revision of that chronology (Slater 1990; Hindle 1990), but few changes to the plan analysis.

Plan analysis is not, on its own, a means of establishing morphological regions, since the patterns of building fabric and land use are not considered. Nor is it a technique for closely dating the plan units that are discerned in a particular town, though a relative chronology of plan units is sometimes reasonably clear. Precise dating relies upon documentary or archaeological evidence and such evidence for the medieval town is often available only for particular sites or properties. However, a careful regionalization of the town plan may allow such specific dates to be applied to wider areas by analogy. Examples of such studies that integrate plan evidence with historical and archaeological evidence are Slater's studies of Hedon, East Yorkshire (1985) and Doncaster, South Yorkshire (1989) and Baker's analysis of Walsall, West Midlands (1989). All were part of detailed archaeological studies and the plan analyses were important in providing wider contextual settings for particular excavation sites and for suggesting sites where future excavation might be able to date elements of the town's expansion.

The broad principles of town-plan analysis in medieval towns have, therefore, been well established over two decades of research by a small group of scholars. However, the lack of a precise, verifiable and repeatable method for plan-unit definition, and still more so for the delimiting of morphological regions, became apparent with the commencement of a major interdisciplinary research project at the University of Birmingham funded by the Leverhulme Trust, which aimed to investigate the interaction of the church and medieval towns. The project began with a pilot study of medieval Worcester which was intended to establish the most suitable sources, techniques and methods for a regional study of the towns of the West Midlands by archaeologists, geographers and historians. The aims of the geographical contribution to this pilot study were, first, to reconstruct the detailed plan of

Worcester as closely as possible to how it might have been in *c.* 1500, and then, second, to examine the morphogenetic development of the city from post-Roman times to the date of that reconstructed plan. It was important that the base map be as accurate as possible, since archaeological data were to be superimposed, and it was important that the techniques for compiling the plan for *c.* 1500 were able to be used in similar studies of other towns. It is the example of the city of Worcester which is utilized in this chapter to explore some of the problems and possibilities of defining plan units and morphological regions in English medieval towns.

COMPILING THE PLAN

For cities of Worcester's size the first accurate plan of sufficiently large scale is almost always the 1:500 Ordnance Survey or Public Health plans of the mid-Victorian period (1886 in Worcester's case). A particular advantage of the 1:500 plans over the 1:2500 scale normally used for small town studies is in the depiction of access points to properties, such as gates and doorsteps, which considerably aids the interpretation of landownership and tenancy units. Of course, the Ordnance Survey was not concerned with units of urban property, but with the accurate depiction of the built environment. None the less, property units can be interpreted relatively easily in less heavily built-up areas. However, in inner urban areas, where building coverage is high and exchanges between tenements may be more rapid and unpredictable, particularly on corner plots, tenancy units can be more difficult to interpret. This is a problem to which Scrase (1989) has recently drawn attention in his study of the medieval city of Wells, Somerset.

A consistent method for the selective retrieval of property-boundary information was, therefore, the first problem to be tackled in establishing a viable way of reconstructing the early plan. The technique used was to show on the base map all divisions between buildings on the street frontages for the distance that they ran back from the frontage without significant deviation or interruption. A maximum dog-leg of *c.* 1.5 metres (nearly 5 feet) actual size was allowed to cater for property boundaries shifting from one side of a boundary to the other. All free-standing boundaries behind frontages were shown, except where they were clearly associated with minor ancillary buildings, and these, too, were extended for the distance they ran between buildings without significant deviation. Some minor frontage divisions between small buildings were subsequently removed to avoid loss of clarity when the map was reproduced at a smaller scale. Also omitted were divisions in what were clearly post-medieval, terrace-cottage infilling of larger containing plots. The method is reconstructed for one street block in figure 3.1.

The second problem was to adopt a logical and consistent approach to the depiction of the extent of the built-up area of the medieval city, given the lack of precise documentary evidence for the period. Here the decision was taken to plot information consistently for all areas shown as occupied at the date of the earliest detailed plan of the city. In the case of Worcester, this is George Young's map of 1779. It seems likely, from comparison with Speed's plans (1610), that the areal extent of cities such as Worcester was not markedly different between 1600 and the later eighteenth century, population growth being accommodated by intensification of building coverage within existing plots. Pre-industrial field patterns and minor suburban service lanes and paths were added from the earliest cartographic sources for the surrounding areas. Finally, the easily recognized industrial-era features were removed from the map, and pre-industrial features restored, through recourse to the city archives and to the deposited plans for railway and canal insertions, road-widening, break-through streets and their associated plot patterns, and municipal services such as market halls and cemeteries.

The recognition and isolation of post-medieval elements of the plot pattern are, of course, not so simple. In Worcester, documentary sources are generally too poor ever to allow a comprehensive plot-by-plot reconstruction of post-medieval evolutionary sequences of the type undertaken, for example, in Ludlow (Lloyd 1978) or Wells (Scrase 1989), and still less the reconstruction of medieval properties attempted for Canterbury by Urry (1967). Again, too few pre-eighteenth-century buildings survive in Worcester, and archaeological excavation has been on too limited a scale, to accomplish the same end by recourse to the physical evidence. That the plan reconstruction of this city is indeed recreating a partly medieval, rather than purely post-medieval, townscape is therefore a matter of probability, rather than certainty. But the level of probability is arguably extremely high given, first, the general case that has been made for the conservatism of plot patterns nationally on theoretical grounds (Conzen 1960, 1962) and, second, the evidence of documentary and archaeological studies (Webster and Cherry 1977; Lloyd 1979). The specific case for the conservatism of plot patterns in Worcester can also be made. The limited documentary research and excavation that have been done suggest this to be so, as does examination of those medieval and sub-medieval buildings that survive in the town, together with research on buildings known from

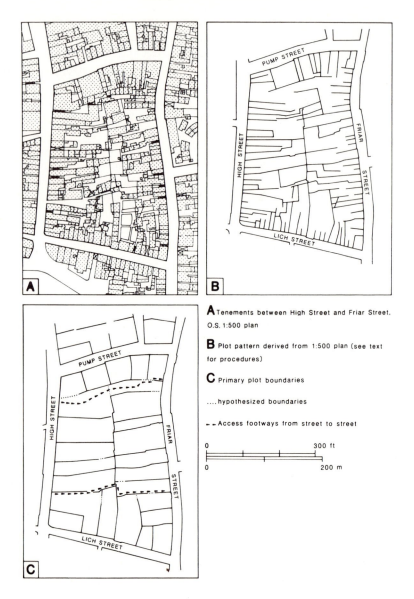

A Tenements between High Street and Friar Street. O.S. 1:500 plan

B Plot pattern derived from 1:500 plan (see text for procedures)

C Primary plot boundaries

.... hypothesized boundaries

-- Access footways from street to street

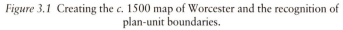

Figure 3.1 Creating the *c.* 1500 map of Worcester and the recognition of plan-unit boundaries.

photographs to have been standing when the large-scale Ordnance Survey plans were made in the 1880s (Hughes and Molyneux 1984; Hughes 1986; Mundy and Dalwood forthcoming). This is not, however, to deny that more work is required to determine, with greater precision, the street-by-street balance between survival, evolution and replacement of buildings and plot patterns in Worcester.

The resultant composite plan of property boundaries represents, as closely as we believe possible for cities with this level of cartographic and documentary source material, the topography of late-medieval Worcester in *c.* 1500 accurately plotted at a scale of 1:500. It is a plan on to which more detailed archaeological, building survey and landownership data can be superimposed. It differs somewhat from the *British Historic Towns* atlases, and from Conzen's analytical plans, in its concentration on plot boundaries. The majority of buildings, with the exception of churches, are omitted in the preparation of this first-stage base map. It is this base map which forms the frame for the subsequent derivation of medieval plan units. For the intra-mural area of Worcester, this base map, and its constituent plan units, are shown in figure 3.2 in a slightly simplified and reduced form.

PLAN-UNIT DERIVATION

Initial impressions of the reconstructed plan of late-medieval Worcester are of extreme complexity. There are few signs of regularity which might suggest organized, large-scale planning; rather, the plan appears to be the product of piecemeal, uncoordinated growth. The street spaces vary greatly in width, straightness and orientation, and the characteristics of the plots lining them are diverse. A cursory glance down the length of the High Street axis of the city (figure 3.2) reveals something of this diversity. The recurrent association between changes in the width and direction of the street, and changes in the character of the plots either side of it, offers an immediate clue to the composite character of the town plan and to the fact that it should be possible to discern plan units which accord with Conzen's original definition.

While recognition of such plan components may be immediate, their accurate resolution raises a number of problems, the first being that of scale, as Conzen himself recognized in his studies of both Alnwick and Ludlow. Areas exhibiting a 'measure of morphological unity' may be defined at very different scales – from the whole intra-mural area down to a minor plot series. Further, in some cases, characteristics which distinguish an area from its neighbours may not be uniformly present

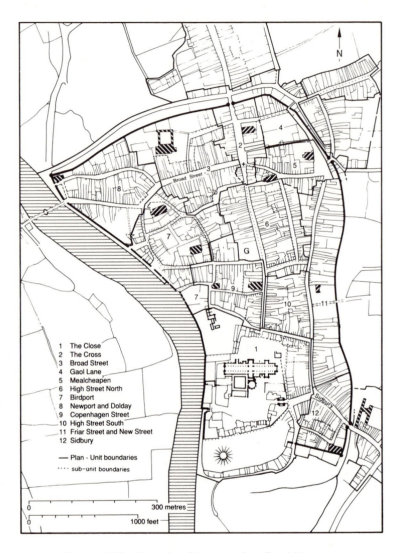

1   The Close
2   The Cross
3   Broad Street
4   Gaol Lane
5   Mealcheapen
6   High Street North
7   Birdport
8   Newport and Dolday
9   Copenhagen Street
10 High Street South
11 Friar Street and New Street
12 Sidbury

— Plan - Unit boundaries
···· sub-unit boundaries

Figure 3.2 The plan units of intra-mural medieval Worcester.

throughout that area. Rather, a core area might exhibit the full range of defining characteristics but have marginal areas around it in which only some are present and where no clear boundary between plan units, or plan seam, is distinguishable (see the discussion of the Birdport plan unit later in this chapter). In the Worcester study, the scale of plan-unit definition was determined as that of the individual street or street block. Smaller-scale variations were isolated and described as sub-units of the larger street-block units, following the practice introduced by Slater (1989) in his analysis of Doncaster. They conform reasonably to Conzen's plan units of second and third orders of magnitude in Alnwick. In morphographic terms, therefore, the plan units defined in Worcester represent the area occupied by plots associated with a street or streets where the orientation, shape and dimensions of those plots and street(s) can be clearly differentiated from their neighbours and have one or more internal characteristics in common. Function can sometimes be taken as a defining characteristic where it is apparent but, for the medieval period, this is unusual.

The street block shown in figure 3.1 provides a simple example of such plan-unit definition. Whereas the plots fronting High Street and Friar Street have similar east–west orientation and similar widths and depths (indeed, a number of boundaries run through the back fence which divides the two plot series), the plots fronting Pump Street are only one-third the depth and of variable width. Further, where the Pump Street plots meet the High Street series, a different orientation is clearly apparent. These differences suggest that a plan seam runs along the rear of the Pump Street plots (where there is also an irregular through footway). In contrast, the plots fronting Lich Street, to the south, have characteristics that suggest they are derivative from the east–west plots of High Street and Friar Street. First, the back-fence line continues through to Lich Street itself; second, there are a number of L-shaped plots with access to both Lich Street and High Street or Friar Street, suggestive of an early period of common ownership; and, third, the plot pattern at the two street corners is one of progressively smaller, interlocking, rectangular plots reflective of a process of successive subdivision of earlier, larger properties. This is perhaps made clearer when the primary boundaries, those defined as running continuously from street front to back fence without significant deviation (Slater 1981), are extracted from the still complex plot pattern shown in figure 3.1B. Such boundaries are shown in figure 3.1C. The Lich Street plot series could be defined as a third-order plan unit but, in this instance, it has been judged that the plots are most likely derivative from the main plot series of High Street and Friar Street and therefore they have not been differentiated. Streets form the plan-unit boundary to the east and south. Eastwards, the plot series on the east side of Friar Street are longer than those in the plan unit concerned, while south of Lich Street they are very much shorter and form part of the cathedral close (discussed later in this chapter). To the west, the boundary also runs along the street (High Street), though the varied plot series of the adjacent plan unit (figure 3.2: plan unit 9) makes this a matter of finer judgement since some of them are of similar dimensions to the plots on the east side of High Street. This is, in sum, one of the most straightforward of all the city-centre plan units to define. The complexity of defining some of the other plan-unit boundaries will emerge in later examples.

Temporal change poses a particular problem for plan-unit definition. If it is accepted that the plot pattern recorded in the nineteenth century is likely to represent a modified, evolved, version of the plot pattern at the end of the Middle Ages (itself, of course, the product of evolutionary change), then it follows that the plan units, too, will reflect such change. In particular, any plan-unit boundaries following plot boundaries rather than streets may have changed as plots on the boundaries underwent the usual processes of mediation, repletion, amalgamation and extension. For example, the plan seam between the High Street and Birdport plan units (figure 3.2: plan units 6 and 7) has a distinctive staggered appearance suggestive of very fluid landownership in the area and the consequently unpredictable exchange of ground between the two plot series. A parish boundary here followed a similar, but separate, staggered north–south line and appears to have represented an earlier junction between the plot series. Excavation on the line of the parish boundary at its most westerly point demonstrated that, on that site, the property boundary it followed was not established until the fifteenth century (Mundy and Dalwood forthcoming). Before that, spreads of industrial residue stretching continuously from the Birdport end of the plots suggested that the boundary between the west- and east-facing plots lay closer to the High Street, as it did later. In other words the seam between the two plan units moved over time as plots were extended or truncated at the rear, the parish boundary representing a tide-mark left by this process. However, as these movements are unpredictable, and unrecoverable without further excavation, the boundary to the plan units shown in figure 3.2 is drawn in its eighteenth/nineteenth-century state and no attempt at reconstruction has been made. Similar difficulties are posed when plan-unit boundaries are drawn through corner plots where rapid changes in medieval plot boundaries have been amply demonstrated by Scrase (1989).

The plan units of Powick Lane, located between the Broad Street (unit 3) and Birdport and High Street (units 6 and 7) (figure 3.2) provide a further example of changes in the outline of plan units. By the 1880s this lane was flanked by small built-up plots and, at that time, arguably constituted a plan unit in its own right. However, the continuance of a number of plot boundaries from the Broad Street frontage through to the lane suggested that the latter's plots were derivative, that is they had been alienated from the primary Broad Street plots. Documentary evidence and archaeological evidence (Currie forthcoming; Mundy and Dalwood forthcoming) support this, and date the start of the process to the end of the Middle Ages. Given the relative certainty of this evidence, in pursuit of a period-specific analysis, the southern boundary of the Broad Street plan unit has therefore been reconstructed on Powick Lane and the derivative, secondary plots belonging to the post-medieval plan unit have been removed. The process observed on Powick Lane almost duplicates that described by Conzen as occurring between Market Street and Green Batt in Alnwick (Conzen 1960: 65–9), though with little sign in Worcester of the development of medial plots. The message is, however, the same: plan units were not, and indeed are not, static phenomena. They are fluid; liable to change in extent as well as in their internal character; and their analysis and cartographic representation therefore demand a consistent and explicit approach to the problems of reconstruction and chronology.

PARISH BOUNDARIES

In attempting to resolve the town plan into its constituent parts, another question that must be faced is how far it is appropriate (if at all) to use non-morphological criteria. Land-use patterns are one example which, despite their inclusion by Conzen as part of the physical make-up of towns, are not specifically plan characteristics, though clearly certain land uses have distinctive building or access requirements which will be reflected in the buildings, plots and locations they occupy within the urban fabric. Another such example is the use of secular and ecclesiastical boundaries. It could be argued that boundaries have both plan form (in terms of the linear symbols used to denote them) and a physical actuality in the townscape (through boundary markers such as stones and crosses). It could also be argued that parish boundaries represented boundaries to property when they were first demarcated and therefore deserve inclusion in any cadastral analysis. Parish boundaries demarcated areas from which tithe was derived and have, in many cases, been found to derive from secular property units (Rogers 1972: 48; Blair 1991: 109–33; Morris 1988: 168–266). An analysis of the relationship between the city's parochial geography and the mapped and excavated secular landscape shows that, while some parochial divisions recorded on late-eighteenth-century maps clearly reflect the form of the town in the tenth century, there has always been a degree of fluidity. Large-scale boundary change took place sporadically, for example, when new parishes were created, particularly on the urban fringe, or others were amalgamated. Similarly, change occurred in Worcester as a result of reorganization following the construction of the city walls. Small-scale change is likely to have been a constant process, parish boundaries shifting as properties they embraced were subdivided or extended (Baker and Holt forthcoming). It would, therefore, be dangerous to assume that a particular property boundary was necessarily of particular significance as, say, a plan seam, by virtue of a greater antiquity deduced from its coincidence with a parish boundary. This might well be so, but may only be established by careful consideration of its archaeological context (if known) and its place in the wider pattern of ecclesiastical and secular boundaries.

The plan analysis of Worcester includes one case where ecclesiastical boundaries have been used over and above purely morphological evidence in the definition of a plan unit; namely the cathedral close, a feature that is characteristic of most English medieval cathedral cities and especially those where the cathedral was monastic, as it was in Worcester. There is little doubt that such closes entirely fail Conzen's definitional test of 'a measure of morphological unity', consisting as they do of a number of very diverse elements. In Worcester, for example, there is the cathedral building, monastic buildings, canons' housing, lay and monastic cemeteries, secular housing, an area which was part of the Norman castle bailey for a period, and the sites of earlier churches. However, despite this morphological diversity, the cathedral close had a quite distinct legal identity in the later Middle Ages. Its boundaries were closely defined and resolutely defended from at least the fifteenth century to the nineteenth century, and they were expressed physically through a precinct wall and by outward-facing, built-up frontages penetrable only through gated entrances. It is this legal separateness and distinctive boundary which have determined its definition as a plan unit, rather than its morphological homogeneity or even its usage, though different types of secular housing are distinguished as sub-units. Similar definitional problems may occur with other large medieval institutions such as palaces and baronial castles. The castle at Alnwick is an example, but there Conzen, too, defined it as a single plan unit.

INTERPRETING PLAN UNITS

Chronology becomes an even greater problem when the plan units, once defined, have to be interpreted. In Worcester, as in most large towns with complex plans, the sequence in which plan units were established and the date at which they were established are rarely evident from the plan evidence alone, particularly where polyfocal development is a possibility. Documentary evidence is of only limited assistance in Worcester as this city, like many others, achieved its greatest medieval extent before the thirteenth century; the parochial geography may offer some guidance to chronology, though the use of church dedications for dating settlement is problematic to say the least (Everitt 1986).

Archaeological sources (controlled excavation and the monitoring of construction work) may be able to provide dating evidence that is not morphologically specific by demonstrating occupation in an area at a certain date, without reference to the spatial framework in which it took place. It was argued earlier that it might be possible to use such evidence in order to date a specific mapped pattern of streets and boundaries, but this should be done only with the greatest caution. In Worcester, occupation on the north side of Sidbury is archaeologically attested from the late Saxon period (possibly from the ninth century, more probably from the tenth century) onwards (Carver 1980: 165), but the apparent orientation of excavated features of this period in relation to the existing street, rather than the underlying Roman street, is insufficient evidence to prove that the occupation was taking place within elements of the plot-pattern known later in the medieval period.

Specific features of a town plan may, of course, be both archaeologically detectable and datable. Biddle and Hill's (1971) use of coins stratified in sequences of road metalling to date the planned street-grid at Winchester remains a classic case, and early road surfaces have been excavated and dated in, for example, London and Gloucester (Vince 1990: 126–7; Heighway 1984: 367–8). The excavation, identification and dating of urban plot boundaries, as represented by fences, ditches or walls, are relatively commonplace, though partly dependent on the scale of excavation. However, boundaries may not always be directly represented by structures; in Worcester, one mapped property boundary first became apparent in excavation by a discontinuity in pit-digging, another by the differential robbing of a stone building (Mundy and

Dalwood forthcoming). Standing buildings may provide a *terminus ante quem* for associated plot boundaries; helpful perhaps in working out later medieval evolutionary sequences and distinguishing medieval from post-medieval (see Baker *et al.* forthcoming), but there will be relatively few cases where even the oldest standing buildings will approach in date the likely latest date for major medieval urban extensions.

In Worcester, as elsewhere, a proportion of the defined plan units are interpreted as representing planned urban extensions created over a short period of time; others are interpreted as the result of piecemeal development taking place over an unknown period of time. Clearly, in the latter case, a plan unit can only be 'dated' at all by a definition of the period over which the settlement it represents developed; for the early-medieval period, such a conclusion will generally be derived only from the archaeological investigation of an adequate sample of individual sites. However, even 'planned' areas may contain hidden internal chronologies that may complicate the application of archaeologically derived dating evidence. There may, for example, be gaps of unpredictable duration between the construction of one or more new streets, the laying-out of primary plots, the creation of secondary plots, and their eventual take-up and actual occupation. An example of this kind of sequence, based on the plan evidence alone, is Slater's (1990) interpretation of the development of the Broad Street/Mill Street plan unit in Ludlow.

It would be difficult to find archaeological dating evidence for an urban extension more apparently conclusive than the excavation of a series of coins with restricted circulation periods stratified on primary street surfaces, but even these cases only establish a *terminus post quem* for a train of potentially separate events. The simultaneous laying-out of a number of boundaries (in so far as it is possible to establish this from archaeological evidence) has, in several towns, allowed excavators to postulate the creation of 'planned', early-medieval developments. Examples include Coppergate, York (Hall 1984); Saddler Street, Durham (Carver 1979); and Chequer Street, St Albans (Youngs *et al.* 1983: 181–2). If, in cases such as these, the excavated plots were to be identified as components of a more extensive surviving or mapped series forming part of a plan unit representing a 'planned' urban extension, the archaeological evidence could reasonably be held to date the plan unit, but only if the plots in question were identifiably a primary component of the plan unit and not a product of a secondary phase in its internal evolution: for example, the subdivision and intensified occupation of a large primary plot.

It might be felt that such arguments are, to some extent, pointless, given the relative crudity of archaeological dating methods. However, as Vince (1990: 27) has recently pointed out, in London at least, there is a growing prospect of very precise dating for the growth of the late-Saxon built-up area through accumulated dendrochronological dates from waterlogged timbers. Such results may, eventually, be replicated elsewhere, offering the hope that, through the careful integration of different sources of evidence – cartographic, documentary and archaeological – and with archaeological sampling maintained on an adequate scale, the physical expansion of towns in the early-medieval period may yet be quantifiable with a reasonable degree of accuracy.

TOWN PLANNING

Over the past decade, much work has been undertaken on the physical form of small- and medium-sized planned towns in Britain, using the techniques of town-plan and metrological analysis to develop a better understanding of the processes of planning in the high-medieval period. This has built upon the pioneering work of Beresford (1967), who provided a national historical and geographical framework, but his plans concentrated almost exclusively upon street patterns. Slater's work (1987, 1990) on such places as Stratford-upon-Avon, Lichfield and Ludlow has shown the intricate nature of development processes in the initial years of a planned town and has tried to discern and differentiate between the ideal plan in the mind of the town's founder or planner, and the reality of development on the ground.

In Worcester, there is a good example of high-medieval planning in the long northern street suburb of Foregate Street and The Tything. Plots were laid out regularly on either side of the street with a back lane to the rear. Metrological analysis of the plots suggested a regular initial module, though most primary plots had been subsequently subdivided in the same manner as in smaller towns. The plan form and development processes were very similar to those of hundreds of small, planned market boroughs laid out along main roads in twelfth- and thirteenth-century England. The only difference was that Foregate Street and The Tything formed a suburb attached to a much larger town.

For towns which were already developed before the twelfth century, even where deliberate planning has long been an accepted part of their development history, there have been few attempts to understand the nature of this planning beyond the level of the street plan and the street-block pattern. Thus, for example, Biddle and Hill (1971) have drawn attention to the planned nature of many of the Alfredian *burhs* of

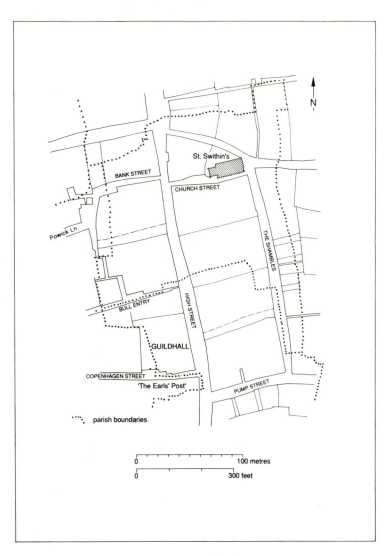

Figure 3.3 The High Street North plan unit (6), showing primary boundaries.

southern England but, even in the case of Winchester, where so much archaeological excavation has been undertaken, the evolution of the town plan continues to be discussed at the level of the street block (Biddle 1976). A similar situation pertains with other towns with large-scale excavation programmes, such as Oxford, Southampton and Canterbury. Crummy (1979) has attempted a metrological analysis of late-Saxon Colchester (Essex) but, again, this is at the level of the street block.

One of the few immediately obvious indications of rectilinear planning in intra-mural Worcester is to be found on the east side of the central section of the axial High Street (figures 3.2 and 3.3). The Shambles, running parallel to the High Street, and two short perpendicular streets (Church Street and Pump Street) define a large rectangular street block. Within this block, as well as on the opposite (west) side of the High Street, were (and to an extent still are) two series of very irregular, elongated plots, those on the east side backing irregularly on to shorter plots facing the Shambles. Much shorter plots, ending on largely continuous back-fence lines parallel to the frontages, lined the east side of the Shambles and the south side of Pump Street. On the west side of the High Street, the majority of the plots backed on to an alley following a markedly staggered course, which has already been described. The area as a whole exhibits a degree of morphological unity in terms of the plot pattern and the arrangement of streets in relation to the High Street, and has been classified as a plan unit (plan unit 6, figure 3.2). The boundaries of the plan unit have been defined by the extent of the distinctive, elongated, irregular plot pattern and by the extent of the dissimilar plots associated with the subsidiary streets.

Amongst the irregular pattern of plots, however, were a number of primary boundaries running straight through from the principal High Street frontage to the subsidiary street or lane to the rear. Metrological analysis suggests that they were spaced at regular intervals of *c.* 156–8 feet (about 46 metres) and appear to represent the boundaries to large, regular primary plots. They confirm the initial impression, based on the rectangularity of the street system alone, that this area consists of the transformed remains of a regularly planned urban layout. There is so far no independent internal dating evidence for the primary features of this landscape. However, there is little doubt that it lay within the defences of the documented late-ninth-century *burh* and was arguably a primary component of it. It represents a planned development of part of the pre-existing axial street leading northwards out of a gate through the old Roman defences. As such, it has contemporary parallels in London and

Wessex (Dyson and Schofield 1984: 298–301; Biddle and Hill 1971). The speed at which these large primary plots were irregularly subdivided is more or less unknown, little archaeological work having taken place within the area. However, the remains of one or more vaulted undercrofts respecting the boundary of one of the secondary plots suggests that the process may well have been advanced by c. 1200 (H. Dalwood, personal communication).

These large rectangular blocks are rather different in scale to the documented primary plots of between one-third and one-half acre (0.1– 0.2 ha) which are typical of thirteenth-century planned new towns (Slater 1981). They are closer in size to layouts in some of the late-developed, fourteenth-century towns and suburbs where initial plots were up to 2 acres (0.8 ha) in size. Such initially large plots suggest the need to provide substantial incentives to the first holders in terms of the potential for subsequent subdivision and subletting. This was so in the mid-fourteenth century, when one of the effects of the Black Death was an oversupply of urban tenements and a fall in potential rental receipts. However, in late-ninth-century Worcester these large plots may have been a means of trying to ensure immediate economic success for this new planned urban extension by spreading the creation of the urban infrastructure amongst a small group of high-status landholders or rural estates and their communities. These large, regular blocks might also be one example of the physical manifestation of an urban *haga*, a term which occurs quite frequently in Anglo-Saxon charters and leases (Hooke 1981, 1990). The word means 'hedged enclosure' in a rural context and may have an equivalent meaning in this urban situation.

Not all *hagae* were of this kind, however, since a well-studied Worcester example can now be precisely located. A charter of 884–901, from Aethelred of Mercia and his wife Aethelflaed to Bishop Waerferth, granted the bishop rights in Worcester which, at his request, they had recently fortified (Sawyer 1968, no. 223). Shortly afterwards, in 904, Waerferth leased a *haga* to Aethelred and Aethelflaed which lay in the north-west corner of the town, in the angle of the defences and the river (Sawyer 1968: no. 1280). Until recently, the defences of the *burh* had not been located and it was therefore not possible to locate the *haga*. However, now that they have been proved to run between Grope Lane and Broad Street, the *haga* may be determined with some accuracy. The lease locates it in the angle of the Severn and the north wall of the *burh*, and thus in the general area of plan unit 7 (figure 3.2). It gives measurements expressed in rods for its north, east and south sides (figure 3.4). These correspond closely with the Grope Lane–Birdport–Copenhagen

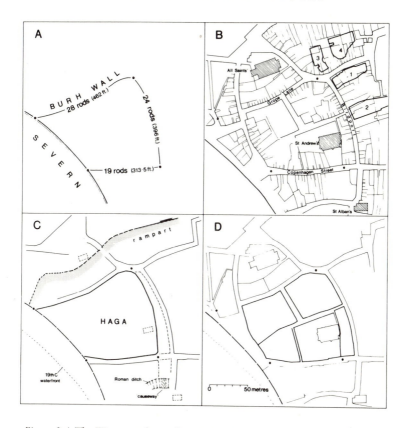

Figure 3.4 The Worcester *haga* of AD 904. A: Documentary evidence: the base of 904. B: The eighteenth- and nineteenth-century plot pattern. C: The 904 *haga* reconstructed. D: Probable initial subdivisions of the *haga*.

Street block if, first, the rods are assumed to be equivalent to the statute perch of 16.5 feet (5.03 metres) and, second, a degree of movement is assumed in the position of the waterfront and in the line of the Birdport towards its southern end. A westward shift in the waterfront, away from the bottom of the slope, may well have taken place through the familiar process of advance by successive revetment and reclamation. A slight eastward shift in the course of the southern end of the Birdport route can be suggested on a number of grounds: the siting of St Alban's church in relation to Little Fish Street – Birdport's southern continuation; archaeological evidence for a brushwood structure interpreted as a causeway crossing the Roman ditch a few metres to the west of the

61

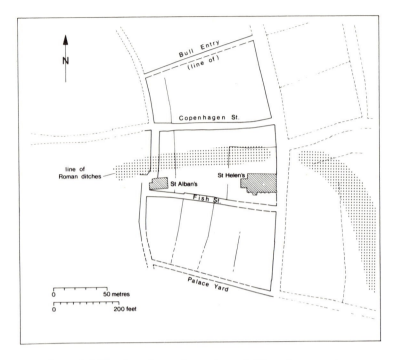

Figure 3.5 The Copenhagen Street plan unit (9): an interpretative reconstruction of the planned area.

street; and the observation of features (in section) interpreted as hollow-ways caused by north–south traffic similarly just west of Little Fish Street (Gelling 1958; Richardson and Ewence 1961).

How long this *haga* had been in existence prior to its only appearance in the documentary record is unknown. Its role was probably primarily commercial, given its riverside location. The better-documented commercial role of the London *hagae* gives some indication of probable developments. There, a *haga* was bought by Bishop Ealhhun in 857; another, adjacent to the waterfront, was granted to Waerferth himself in 889 (Holt forthcoming; Dyson 1978). The finding of a *sceat*, an Anglo-Saxon coin, in excavations immediately to the east of the Worcester *haga* suggests the possibility that the riverside there may have been commercially exploited as early as the eighth century (Mundy and Dalwood forthcoming). The fate of the bishop's *haga* after 904 is not recorded, but the plan evidence suggests that it was subdivided into four smaller enclosures, one of which contained St Andrew's church. Each of these was later partly colonized by more conventional urban plots (figure 3.4), but the plan form could not have been more different to the High Street area of plan unit 6. Here the townscape is dominated by narrow streets, the plots often being ill defined away from the frontages and the whole being morphologically diverse. It is that diversity which is this plan unit's principal characteristic, distinguishing it from those areas around. The extent to which it is a planned development is difficult to ascertain, but planning seems limited to the primary subdivision of the four street blocks. It is also difficult to suggest any firm chronology for the development of this area precisely because of its irregularity.

By way of contrast, a final example of early-medieval planning is provided by plan unit 9 (figures 3.2 and 3.5) which can be dated reasonably precisely. This Copenhagen Street plan unit is characterized by a framework of east–west thoroughfares leading off the west side of the High Street, associated with north–south plots, several boundaries of which passed from one street frontage through to another. St Alban's church, at the west end of Fish Street, and St Helen's at the east end, are both documented in 721 when they were granted to Evesham Abbey (Barker 1968–9). The general impression is that the whole area was subject to a degree of regular planning. If allowance is made for a northward deflection of the east end of Fish Street (probably representing encroachment into an open space around St Helen's church), and for the northward migration of part of Palace Yard (the southernmost street) as a result of encroachment by the bishop's palace precinct (whose pre-thirteenth-century ranges all lie to the south), the whole area can be reconstructed as a simple grid layout of three street blocks of equal area, distorted to accommodate the curve of the High Street and the natural topography. The junctions of each of the east–west thoroughfares are spaced at regular intervals (74 metres or about 240 feet) along the High Street.

Archaeological evidence offers a transforming perspective on this hypothesis. Both this Copenhagen Street plan unit and the neighbouring plan unit 10 (figure 3.1) overlie the earthwork defences of the Roman town (Gelling 1958; Barker 1968–9: figures 4–6). The defences were substantial: the ditch was 27 metres (about 90 feet) wide, and the rampart within it must have been of comparable size. If such an obstacle had remained in place when these areas were first built up, it must inevitably have left some trace in the visible or mapped landscape – by influencing the formation of property boundaries, for example. Yet the line of the Roman defences is almost completely unrepresented in the landscape of either plan unit, suggesting that the defences had been levelled by the time development took place. Deliberate levelling of the defences had been postulated from the evidence of the ditch excavated in the 1950s. The primary slow silting and entrance causeway had been covered by 'back-filling containing layers of marl, sand, loam and slag, with Roman pottery' (Carver 1980: 302; Gelling 1958). The levelling of the defences might, most plausibly, be seen as an immediate prelude to redevelopment of the area, the 'groundworks' in modern contractors' terminology: arguably the urban development represented by the plan units under discussion. Whether this preceded the creation of the *burh* in the late ninth century or, perhaps more probably, succeeded it, the *burh* defences rendering this stretch of the Roman defences redundant, are impossible to prove at present but, clearly, this area of the city represents a zone of late-Saxon town planning which must have been reliant on higher-order decision-making.

CONCLUSIONS

It could be argued that this chapter has said little about morphological regions in medieval towns. Rather, it has concentrated upon the methods of analysing the complex plans of towns with a long developmental history, such as the city of Worcester. However, it is inevitable that any discussion of medieval urban morphological regions will concentrate upon the town plan because the evidence for building fabric and land use is usually fragmentary or non-existent. Even where it is possible to recover some evidence for these aspects of the medieval townscape, it can usually only be done with the resources of a major interdisciplinary research team such as that which was assembled for the analysis of the archaeological excavations in medieval Winchester (Biddle, 1976; Keene 1985). Inevitable, too, is a reliance upon the evidence accruing from archaeology. In cities of the size of Worcester, which have continued to expand in the industrial and modern periods, there are rarely many domestic-scale medieval buildings surviving and even institutional buildings such as churches will normally have undergone several phases of alteration, enlargement or redevelopment. The town plan, too, will have been adapted and transformed in a multiplicity of ways but, as Conzen's work suggested thirty years ago, the town plan is the most resistant to change of the three systematic form complexes used in the definition of morphological regions and it is therefore sensible to begin any study in the regionalization of a medieval townscape with an analysis of the plan. Consequently, this chapter has attempted to specify, as clearly we can, consistent methods for reconstructing the late-medieval plan of a city from first-edition, large-scale plans of the nineteenth century, and for the subsequent regionalization of that plan into plan units.

Once this has been done, evidence from the historical record, from built fabric survival and from archaeological excavation, can be added to the plan. Even where such evidence is as fragmentary as it is in Worcester, a firm topographic base plan can suggest new interpretations of well-known evidence. Thus, in Worcester, the 904 *haga* has been subjected to much study but plan and archaeological evidence now enable this to be located in the developing townscape and to be characterized as an irregularly developed riverside zone of narrow streets and small plots which was probably the location for a variety of craft industries, warehouses and poorer domestic tenements. In other words, it was a distinctive morphological region (in the proper sense of that term) in the townscape of medieval Worcester. The remaining three areas of the early-medieval town which have been considered in detail in this chapter can also be characterized in this broader sense, and appropriately differentiated from each other.

Much work remains to be done in Worcester. However, the archaeological excavations undertaken already allow a crude relative and absolute chronology to be determined for the regions of the town. It is the integration of this archaeological dimension with the plan analysis which illuminates so much more of the medieval townscape. The example of the redevelopment of the Roman defences in the Copenhagen Street region shows this as clearly as anything. Analysis of historical data will allow further information to be added on the social and economic characteristics of the morphological regions of the city, though, again, Worcester is less well off in this respect than many other towns.

Morphological regions of medieval towns can be determined with an acceptable level of accuracy, therefore, but we cannot pretend it is other than a labour-intensive and time-consuming exercise. It can, however, be extremely rewarding in transforming our understanding of the physical nature of important cities in the medieval world.

REFERENCES

Baker, N. J. (1989) *The Archaeology of Walsall*, Birmingham: Birmingham University Field Archaeology Unit.

Baker, N. J. and Holt, R. A. (forthcoming) *The Church and Urban Growth: Worcester and Gloucester*, Leicester: Leicester University Press.

Baker, N. J., Lawson, J. B., Maxwell, R. and Smith, J. T. (forthcoming) 'Further work on Pride Hill, Shrewsbury', *Transactions of the Shropshire Historical and Archaeological Society* 68.

Barker, P. A. (ed.) (1968–9) 'The origins of Worcester', *Transactions of the Worcestershire Archaeological Society*, 3rd series, 2.

Beardsmore, C. (1980) 'Documentary evidence for the history of Worcester city defences', in Carver, M. O. H. (ed.) 'Medieval Worcester', *Transactions of the Worcestershire Archaeological Society*, 3rd series, 7: 53–64.

Beresford, M. W. (1967) *New Towns of the Middle Ages*, London: Lutterworth.

Biddle, M. (ed.) (1976) *Winchester in the Early Middle Ages*, Oxford: Clarendon Press.

Biddle, M. and Hill, D. (1971) 'Late Saxon planned towns', *Antiquaries Journal* 51: 70–85.

Blair, J. (1991) *Early Medieval Surrey*, Stroud: Alan Sutton.

Bond, C. J. (1990) 'Central place and medieval new town: the origins of Thame, Oxfordshire', in Slater, T. R. (ed.) *The Built Form of Western Cities*, Leicester: Leicester University Press.

Bradley, J. (1985) 'Planned Anglo-Norman towns in Ireland', in Clarke, H. B. and Simms, A. (eds) *The Comparative History of Urban Origins in Non-Roman Europe*, BAR International Series 255, Oxford: British Archaeological Reports.

Bradley, J. (1990) 'The role of town-plan analysis in the study of the medieval Irish town', in Slater, T. R. (ed.) *The Built Form of Western Cities*, Leicester: Leicester University Press.

Carver, M. O. H. (1979) 'Three Saxon tenements in Durham city', *Medieval Archaeology* 23: 1–80.

Carver, M. O. H. (1980) 'Medieval Worcester', *Transactions of the Worcestershire Archaeological Society*, 3rd Series, 7.

Clarke, H. B. and Dyer, C. C. (1968–9) 'Anglo-Saxon and early Norman Worcester: the documentary evidence', in Barker, P. A. (ed.) 'The origins of Worcester', *Transactions of the Worcester Archaeological Society*, 3rd series, 2: 27–33.

Conzen, M. R. G. (1960) *Alnwick, Northumberland: a Study in Town-Plan Analysis*, Publication 27, London: Institute of British Geographers.

Conzen, M. R. G. (1962) 'The plan analysis of an English city centre', in Norborg, K. (ed.) 'Proceedings of the IGU Symposium in Urban Geography, Lund, 1960', *Lund Studies in Geography Series B*, 24.

Conzen, M. R. G. (1968) 'The use of town plans in the study of urban history', in Dyos, H. J. (ed.) *The Study of Urban History*, London: Edward Arnold.

Conzen, M. R. G. (1988) 'Morphogenesis, morphological regions and secular human agency in the historic townscape, as exemplified by Ludlow', in Denecke, D. and Shaw, G. (eds) *Urban Historical Geography: Recent Progress in Britain and Germany*, Cambridge: Cambridge University Press.

Crummy, P. (1979) 'The system of measurement used in town planning from the ninth to thirteenth centuries', in *Anglo-Saxon Studies in Archaeology and History*, I, BAR British Series 72, Oxford: British Archaeological Reports.

Currie, C. K. (forthcoming) 'Historical background' in Mundy, C. F. and Dalwood, H. (eds) *Excavations at Deansway*, London: Council for British Archaeology.

Dyson, T. (1978) 'Two Saxon land grants for Queenhithe', in Bird, J., Chapman, H. and Clark, J. (eds) 'Collectanea Londoniensia: Studies presented to R. Merrifield', *London and Middlesex Archaeological Society*, Special Paper 2.

Dyson, T. and Schofield, J. (1984) 'Saxon London', in Haslam, J. (ed.) *Anglo-Saxon Towns in Southern England*, Chichester: Phillimore.

Everitt, A. (1986) *Continuity and Colonisation*, Leicester: Leicester University Press.

Geisler, W. (1924) *Die Deutsche Stadt*, Stuttgart.

Gelling, P. (1958) 'Excavations by Little Fish Street', *Transactions of the Worcestershire Archaeological Society*, NS 35: 67–70.

Hall, R. (1984) *The Viking Dig*, London: Bodley Head.

Haslam, J. (ed.) (1985) *Anglo-Saxon Towns in Southern England*, Chichester: Phillimore.

Heighway, C. (1984) 'Saxon Gloucester', in Haslam, J. (ed.) *Anglo-Saxon Towns in Southern England*, Chichester: Phillimore.

Hindle, B. P. (1990) *Medieval Town Plans*, Princes Risborough: Shire Publications.

Holt, R. A. (forthcoming) *The Church and Urban Society in the Later Middle Ages*, Leicester: Leicester University Press.

Hooke, D. (1981) *Anglo-Saxon Landscapes of the West Midlands: the Charter Evidence*, BAR British Series 95, Oxford: British Archaeological Reports.

Hooke, D. (1990) *Worcestershire Anglo-Saxon Charter Bounds*, Woodbridge: Boydell Press.

Hughes, P. (1986) *Worcester Streets: Blackfriars*, Worcester: privately printed.

Hughes, P. and Molyneux, N. A. D. (1984) *Worcester Streets: Friar Street*, Worcester: privately printed.

Keene, D. (1985) *Survey of Medieval Winchester*, Oxford: Clarendon Press.

Lloyd, D. (1979) *Broad Street, its Houses and Residents through Eight Centuries*, Ludlow Research Paper 3, Birmingham: Ludlow Historical Research Group.

Lobel, M. D. (ed.) (1968) *Historic Towns I*, London: Scolar Press.

Lobel, M. D. (ed.) (1975) *Historic Towns II*, London: Scolar Press.

Lobel, M. D. (ed.) (1990) *Historic Towns III: the City of London*, Oxford: Oxford University Press.

Morris, R. (1988) *Churches in the Landscape*, London: Dent.

Mundy, C. F. and Dalwood, H. (forthcoming) *Excavations at Deansway, Worcester 1988–9*, London: Council for British Archaeology.

Richardson, L. and Ewence, P. F. (1961) 'City of Worcester College for Further Education: its geology and archaeology', *Transactions of the Worcestershire Naturalists Club* 11, 4 (for 1960–1): 226–34.

Rogers, A. (1972) 'Parish boundaries and urban history: two case studies', *Journal of the British Archaeological Association*, 3rd series, 35: 46–64.

Sawyer, P. H. (1968) *Anglo-Saxon Charters: An Annotated List and Bibliography*, London: Royal Historical Society.

Scrase, A. J. (1989) 'Development and change in burgage plots: the example of Wells', *Journal of Historical Geography* 15, 4: 349–65.

Simms, A. (1990) 'Medieval Dublin in a European context: from proto-town to chartered town', in Clarke, H. (ed.) *Medieval Dublin: the Making of a Metropolis*, Blackrock: Irish Academic Press.

Slater, T. R. (1981) 'The analysis of burgage patterns in medieval towns', *Area* 13, 3: 211–16.

Slater, T. R. (1985) 'Medieval new town and port: a plan-analysis of Hedon, East Yorkshire', *Yorkshire Archaeological Journal* 57: 23–51.

Slater, T. R. (1987) 'Ideal and reality in English episcopal medieval town planning', *Transactions, Institute of British Geographers* NS 12, 2: 191–203.

Slater, T. R. (1988) 'English medieval town planning', in Denecke, D. and Shaw, G. (eds) *Urban Historical Geography: Recent Progress in Britain and Germany*, Cambridge: Cambridge University Press.

Slater, T. R. (1989) 'Doncaster's town plan: an analysis', in Buckland, P. C., Magilton, J. R. and Hayfield, C. (eds) *The Archaeology of Doncaster, 2 – the Medieval and Later Town*, BAR British Series 202, Oxford: British Archaeological Reports.

Slater, T. R. (1990) 'English medieval new towns with composite plans', in Slater, T. R. (ed.) *The Built Form of Western Cities*, Leicester: Leicester University Press.

Slater, T. R. and Lilley, K. D. (forthcoming) 'The British Historic Towns Atlas: a critique and international comparison', *Urban History*.

Spearman, R. M. (1988) 'The medieval townscape of Perth', in Lynch, M., Spearman, M. and Stell, G. (eds) *The Scottish Medieval Town*, Edinburgh: John Donald.

Strahm, H. (1935) *Studien zur Gründungsgeschichte der Stadt Bern*, Bern: Francke.

Urry, W. (1967) *Canterbury under the Angevin Kings*, Historical Studies 19, London: Athlone Press.

Vince, A. (1990) *Saxon London: an archaeological investigation*, London: Seaby.

Webster, L. E. and Cherry, J. (eds) (1977) 'Medieval Britain in 1976', *Medieval Archaeology* 21: 248–9.

Youngs, S. M., Clark, J. and Barry, T. B. (1983) 'Medieval Britain in 1982', *Medieval Archaeology* 27: 271–3.

4

PALACES AND THE STREET IN LATE-MEDIEVAL AND RENAISSANCE ITALY[1]

David Friedman

INTRODUCTION

The relationship between the private house and the environment of public space in the Italian city underwent a fundamental reordering in the late Middle Ages. The catalysts were, simultaneously, the new prominence given to the street as an instrument of spatial organization by the merchant–artisan regimes that gained control of the state in this period and the monumentalization of the private residence by builders from the class of men that formed the government. Despite the fact that the officials who commissioned the new streets and the men who raised palaces were sometimes the same people, the two urban types did not, at first, enjoy an untroubled relationship. The new ruling class discovered the ideal form of the street well before they were willing, as individuals, to give up some traditional privileges associated with property ownership that contradicted it. It is not until the Renaissance that street and palace – and this statement is also true for modest domestic architecture – were set into the more or less symbiotic relationship in which they continued until the twentieth century.

STREETS AND URBAN GROWTH

The late Middle Ages was a period of spectacular urban growth throughout Italy. The city of Florence, for example, began a circuit of walls in 1284 that expanded the area of the city five-fold. The newly enclosed land was developed co-operatively by its owners and by the government. In all cases development was based on streets. The new streets were public streets, and as such only one of a variety of passages through the city. Distinctions were both physical and legal. A *via*

69

vicinale, or neighbourhood street, was narrower than a public one. The 1342 statutes of Perugia set a minimum width of 8 feet (about 2.4 metres) for neighbourhood streets and 10 feet (about 3 metres) for public ones. A public street, the priors wrote, is one where 'anyone may pass and which leads to a gate in the walls (*uscita*)' while a *via vicinale* goes only to 'certain places, and has no exit' (Degli Azzi 1913–16: book IV, rubric 18; Grohmann 1981: 51). The commune exercised authority here only with the approval of the neighbours.[2] A third kind of street, *via privatorum*, enters the public record only when its owners made use of the city's surveyors to place *termini* marking boundaries.[3]

The 'public street' was the real theatre of government activity. Among the first acts of new communes was an assessment of their physical estate. The 1210–22 statutes of Volterra require men who had usurped public land or diminished the '*viam publicam in civitate Vulterre vel plateas comunis*' to acknowledge it to the consuls and *podesta* (Fiumi 1951a: 45, from Statute G.3 [1210–22] book I, rubric 90). Later in their histories, communes maintained a systematic vigil against encroachment and even appointed special officers for this task. In Viterbo the *balivi viarum* decided questions between neighbours, set *termini*, cleared streets of obstructions, and commanded abutters to repair the street (Ciampi 1872: 465, 468–9, 478, from the statutes of 1251–2, book I, rubrics 47, 64 and 97).[4] At the beginning of his six-month term of office, the Florentine *Capitano del Popolo* was required to send messengers through the city and territory demanding that anyone occupying any part of a public thoroughfare restore it to the public domain and destroy any offending construction. The *Capitano* employed a building master, a surveyor and an assistant to carry out surveillance on the site and two messengers to announce the survey. After a month's period of grace, local officials in both the city's parishes and the townships of the countryside became responsible for reporting offenders to the *Capitano* (Caggese 1910: 177–8, book IV, rubric 8).

The picture of the city painted in legislation like this suggests a fundamental tension between the government and men of property. This was not the case. In the Italian commune, where legislators and homeowners were the same people, the act of house building was explicitly seen as a virtue, the sign of full participation in civic life. Property ownership was even a requirement of citizenship. The barons from whom the communes won their territory in the countryside were 'pacified' by forcing them to live in the city for a certain part of each year and to own a residence there. When they returned at intervals to their country estates, the house remained as a hostage to the city. Even willing immigrants, at least until the mid-thirteenth century in Viterbo, were required to buy a house or large lot within one month of swearing citizenship (Ciampi 1872: 520, from the statutes of 1251–2, book III, rubric 102). Newcomers at San Gimignano in the same period received a building lot measuring 24 by 12 *braccia* (about 15 by 7 metres) from the commune and five years of tax freedom with the obligation to build a house within six months (Pecori 1853: 722–3, from the statutes of 1255, book IV, rubric 18).[5] Additional legislation, which allowed any house builder to acquire unoccupied lots at a fixed price, identified construction as the commune's priority (Pecori 1853: 734, from the statutes of 1255, book IV, rubric 74).[6] The commune recognized a responsibility towards those who invested in property in the city by underwriting public fire insurance for these buildings. At San Gimignano, Siena and Florence, houses on the tax rolls that were damaged by fire were repaired at public expense, and the crime of arson was punished by execution (Pecori 1853: 683, from the statutes of San Gimignano, 1255, book I, rubric 46; Lisini 1903: 302, (book I, rubric 465); Caggese 1921: 74, (book I, rubric 28)).

Homeowners who were identified with their residences also had obligations to the area surrounding it. The people who in some places in Italy still sweep the streets in front of their house continue a practice that was first recorded in law in the mid-thirteenth century. The San Gimignano regulation required that house occupants sweep the paving in front of their property every Friday evening or face a fine of 12 *denari* (Pecori 1853: 676, from the statutes of 1255, book I, rubric 35). In Siena 'the people who use the street,' that is, the neighbours, paid to pave it, and a representative of the *contrada*, serving as district *provisores lastrichi*, supervised repair (Lisini 1903: II, 40–1 (book III, rubric 71)). In Viterbo the *balivi viarum* divided the expenses for paving a street in the unbuilt area inside the town wall between the owners of all the gardens, whether they abutted it or not, to which the street gave access (Ciampi 1872, from the statutes of 1251, book IV, rubric 120). Even relatively large projects were handled in the neighbourhoods. In 1297, for example, when the Sienese commune wanted to improve an area of town considered 'dark and narrow', it was the people who owned property in the parish who paid for the houses that were destroyed to create a piazza. Individual contributions were assessed by a committee of six assembled by the commune from men of the parish, and the work was overseen by the communal street officers (Lisini 1903: II, 123 (book III, rubric 272, dated May 1297)). Local financing was based on the principle of 'those who benefit most, pay most'.[7] In return

Figure 4.1 Borgo Ognisanti, Florence: street surveyed in 1279 (photograph by the author, 1991).

for shouldering the lion's share of the responsibilities of communal life, householders received privileges that significantly modified the character of the street and influenced the form of domestic architecture throughout the late Middle Ages.

In its ideal form, the public street of the late Middle Ages was physically regular in every way conceivable to its builders (figure 4.1). It was straight, laid out *dritta corda* and *recta linea* over as long a distance as the terrain and previous building allowed (Pampaloni 1973: 114–15, document 66, dated 7 and 24 January 1298). It was paved and graded, and was meant to be enclosed by continuous rows of buildings. In Florence, garden walls 8 feet (about 2.4 metres) tall were required on the street front of vacant lots (Caggese 1921: 353 (book IV, rubric 69)). According to the legislation for their construction, streets disciplined in this way were considered 'useful and proper and beautiful' – they 'honoured' the city.

Except in areas developed on the urban periphery, the streets of cities, at least until the later fourteenth century, fell far short of this ideal. The old streets in the centres of towns, where the best buildings were sited and activity was most intense, were not regular enough, in part because of the privileges conceded to property owners and house-builders by the government.

STREETS AND PRIVILEGE:
THE RIGHT TO ENCROACH

Encroachments at street level

What the government gave was the use of a certain amount of public space immediately adjacent to every piece of private property. Legislation names the elements of buildings that might borrow public space and defines the terms of use. Ambrogio Lorenzetti's *Good Government* fresco in the Palazzo Pubblico in Siena of 1338–9 illustrates the street furniture used for work and display by artisans' shops (figure 4.2). Window counters (*finestre*), tables (*deschi*), benches (*banchi* or *panche*), stools (*predelle*), as well as storage vessels, tools and merchandise all occupy a part of the street. The fixtures were of wooden construction, which meant that they were movable (in fact many of them were taken in each night) and distinct from the masonry structure of the building. The legislation was usually written in the form of a prohibition, which asserted the commune's authority over the space, and a concession: for example, in a Florentine law of 1325, 'no one shall place tables or benches outside his shop in the Via Calimala from the old market to the Ponte Vecchio, unless they are no wider than one *braccia* [about 60cm]' (Caggese 1921: 356–7 (book IV, rubric 69)).[8] In Siena, the government conceded a zone twice as wide and allowed property owners to erect an unspecified range of wooden structures, including poles to support straw mats that served as awnings (Lisini 1903: II, 47–8 (book III, rubric 88)).[9] In this case, the privileges are expressly tied to property ownership. Only on the Campo, the market square, could anyone raise a free-standing awning over temporary tables (Lisini 1903: II, 29–30 (book III, rubric 38)).[10] Everywhere else in the town, awnings and tables had to be attached to a house or shop. Movable tables or carts, potentially so disruptive to the social and commercial order fixed in property, were explicitly banned (Lisini 1903: II, 29–30 (book III, rubric 39)).[11]

There was one appendage to the house with which communal regimes were less lenient. The Sienese statutes not only prohibit the construction of stairs occupying the public street, but also demand their removal. According to the legislation some of these were made of wood, like the *finestra*, but that seems to have mattered less than the fact that, perhaps because of their permanence, they posed a greater obstacle to viability (ASS, *Statuti* 26 [1337], c. 254r., rubric 353).[12] The 1251 statutes of Viterbo, written on the eve of the papacy's relocation to that city and at the height of its power in northern Lazio, prohibit stairs built

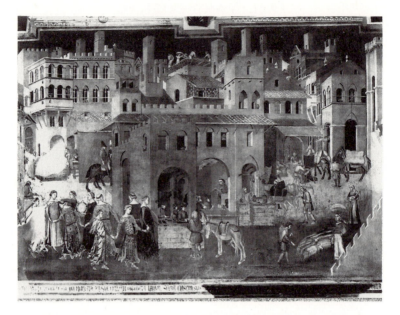

Figure 4.2 Good Government in the City. Detail of a picture by Ambrogio Lorenzetti, 1338–9, in the Palazzo Pubblico, Siena (photograph by the Soprintendenza BAS–Siena: negative 38827).

in the street for this reason, but also because they 'damage the status and appearance of the street' (Ciampi 1872: 507, from the statutes of 1251, book I, rubric 39). Despite the unequivocal tone of the law, property holders did not lose this privilege. The most picturesque ornaments of the very well-preserved historic centre of modern Viterbo are the *proferli*, external stairs, that survive there in great numbers.

Jetties and balconies

The upper levels of houses and palaces enjoyed even more licence than the shops at street level. The requirements of traffic constrained communal officers, no matter how well disposed they were to the space needs of their peers, but above the height of the tallest load on the largest pack animal, a height usually set at 10 feet (about 3 metres) or 12 feet (about 3.7 metres), houses were almost completely free to expand into the area above the street. Even when jettied upper storeys began to receive the attention of the government, the standards were very different from those that were applied at ground level. The 1224

Figure 4.3 Bad Government in the City. Detail of a picture by Ambrogio Lorenzetti, 1338–9, in the Palazzo Pubblico, Siena (photograph by the Soprintendenza BAS–Siena: negative 32755).

statutes of Volterra set a 6-*braccia* or 12-foot (about 3.7 metres) minimum height for the bottom of a *ballatoio*, or jettied upper level, facing on to a public street and a limit of 2½ *braccia*, 5 feet (about 1.5 metres), for its projection. The same documents limited the width of benches and tables at street level to three-quarters of a *braccia* (about 45 centimetres) (Fiumi 1951a: I, 180–1, Statute G.1 (1224) rubric 143).

The *Good Government* fresco records a broad spectrum of jettied structures. They range from simple balconies to completely enclosed wooden pavilions outside the main bearing wall. There also appear to be buildings in which the external wall itself is projected out over the street. The supports for the cantilevered walls are heavy timber beams that penetrate the main structural wall, braced from below by diagonal struts or curved branches like those shown by Ambrogio. The wall of the jetty is relatively light. When they are not made of wooden planks, they are of the lightest possible masonry, as thin as 6 inches (about 15 centimetres) (Sanpaolesi 1939: 260). An image in Ambrogio's *Bad Government* pendant to the *Good Government* scene in the Palazzo Pubblico shows wreckers destroying a jettied structure which is revealed to have been built of a light masonry reinforced with timber (figure 4.3).

The medieval documents do not tell what the space created by the jetties was used for. It is only with their destruction that contemporary testimony speaks about this part of the house. The records of the office of the Sienese government that protected the physical environment of the city, the *Ufficiali sopra l'ornato*, preserved for the period between 1444 and 1478, contain the petitions of householders who had been ordered to destroy jetties. Many ask for an appointment to government office to defray the cost of the work which they inevitably claim will be 'a great ornament to the city'. The government's willingness to subsidize these improvements is another sign, not only of its commitment to the urban renewal effort, but also of its support for the property-owning class. In pressing their claims, homeowners give a vivid account of their loss. From these remarks it is clear that the Sienese jetties that survived to the fifteenth century are mostly the kind in which the whole front wall of the building was moved out over the street, as it is in the jetties of timber-framed structures. The space gained by the jetty was integrated into the upper-level rooms where it could represent a significant percentage of the floorspace of the house. One house lost jetties that projected almost 2.5 metres over the street, extending 12 metres on one storey and 6 metres on the other (ASS, *Concistoro* 2125, petition number 15, granted 20 July 1463). The property decreased in value by 200 florins when the jettied section was taken. The owner, Armaleo di conti degli Armalei, claimed that this was a great loss. He shared the house with his mother, and said that without a subsidy he would be driven into poverty and have to leave the city. Another petitioner claimed that the removal of jetties 'diminished the size of the house to such an extent that it ruined the main hall and we are left as if in a hole' (ASS, *Concistoro* 2125, petition number 116, granted 5 September 1472). Another claimed that he had to rebuild because 'without the jetties the house has become uninhabitable' (ASS, *Concistoro* 2125, petition number 24, no date).

When the projections are in the form of galleries, with the main stone load-bearing wall intact from cellar to roof, the space of the jetties is separate from the main rooms and its shape is long and narrow. The niches built into the outer face of the stone wall (figures 4.4 and 4.5) are a physical sign of the integration of the galleries into domestic life from the moment of the construction of the house. Florentine houses still include a number of such spaces. We can gain an idea of how they were used from the petitions of property owners affected by the regulations prohibiting the reconstruction of jetties written by the Medici dukes in

Figure 4.4 House at Via Caprarecce 41–3, Viterbo (photograph by the author, 1991).

the sixteenth century (Salvagemini 1976). A petitioner living on Via della Stufa near S. Lorenzo described a jettied area about 60 by 210 centimetres that had a little study (*un poco di scrittoio*) and a toilet (ASF, *Capitani di Parte Guelfa*, numeri neri 727, petition number 287, 19 February 1572). Another on the Via Larga contained stone stairs, halls and bathrooms (ASF, *Capitani di Parte Guelfa*, numeri neri 727, petition number 75, 27 August 1572). Another, on a cross street of Via Ghibellina, a main entrance route to the centre of town from the east, contained a toilet and a corridor which connected the main part of the house with the owner's bedroom. The petitioner laments that 'without it [the jetty] I will be able to inhabit this house only with great difficulty' (ASF, *Capitani di Parte Guelfa*, numeri neri 711, petition 211, October 1561). A neighbour suffering from the effects of the same renewal project in the Via Ghibellina area reports that a jettied area over an

Figure 4.5 Detail of house at Via Caprarecce 41–3, Viterbo (photograph by the author, 1991).

adjacent alley (he, like his neighbour, had already lost the jetty on the main street) contained a toilet and a chimney (ASF, *Capitani di Parte Guelfa*, numeri neri 711, petition number 197, 7 October 1561).

The cities of Ambrogio's *Good* and *Bad Government*, while full of the most accurate visual description that has survived, are still idealized. The large number of jetties and balconies that he paints may still misrepresent reality. The Sienese statutes of 1309 describe a less orderly situation. In the very heart of the city, the old forms ruled. The Campo itself, the town hall square, was surrounded by palaces covered with balconies and galleries which a law of 1297 hoped, as palaces were rebuilt, to have replaced by multiple-light windows with marble columns, like those on the Palazzo Pubblico and those in Ambrogio's fresco (Lisini 1903: II, 29 (book III, rubric 37)). On the main streets of the city, those leading to the largest churches and to the gates, jetties

Figure 4.6 The Blessed Agostino Novello saves a boy who has fallen from the upper storey of a house (picture by Simone Martini, 1309, in the Pinacoteca Nazionale, Siena: photograph by the Soprintendenza BAS–Siena: negative 30967).

also proliferated. A law to improve the situation demanded only that the streets be 'open to the sky' for one-third of their width (Lisini 1903: II, 17–18 (book III, rubric 5)).[13] The 1309 image by Simone Martini of the Blessed Agostino Novello saving a boy who has fallen from an upper floor gives a good idea of how the jettied galleries could dominate the streetscape (figure 4.6).

In those cities of Italy less strongly marked by post-medieval urban renewal, the surviving thirteenth-century housing stock still preserves traces of the wooden structures that crowded the streetscape and hid the

Figure 4.7 Casatorre Buomparenti, Volterra (photograph by the author, 1991).

load-bearing masonry. The late-twelfth-century double house at 41–3 Via Caprarecce in the Pian Scarano section of Viterbo is a three-storeyed structure with a gable facing the street (figures 4.4 and 4.5). Rows of beam-holes indicate that the building had wooden galleries extending across the entire width of the street wall on two storeys. At the first jettied level, each side of the house had an arched doorway leading to the gallery and a gable-topped niche (like similar niches common on the inside of these houses) for storage. At the base of the roof gable is a row of lightly projecting stones which were probably the coping of a pent roof that protected the wooden structures. Except for the gable and the wall at ground level with its decorated windows, the street front of this building consisted primarily of a wooden screen (Andrews 1982).[14]

Wooden superstructures also dominated the residential architecture of medieval Volterra, and here examples survive that were built for merchants and noblemen as well as artisans. The robust tower house of the head of the powerful Buomparenti clan (figure 4.7), mentioned in documents of the 1210s and 1220s, and probably built in the second half of the preceding century (Fiumi 1951b: 102–3), has beam-holes and stone corbels that mark three levels of galleries. Hook-shaped stone brackets also survive that once held the ridge beam of wooden pent roofs to protect the galleries and the openings into rooms at ground level. Doors lead out to where the galleries once hung, and windows with elaborately carved tracery also survive. Because they would barely have been visible behind the wooden superstructure, the windows must have as their primary function the decoration of the interior rather than the wall on the street. The top of this tower house was covered by a raised roof whose wooden supports were fixed to vertical stone collars on the exterior walls.[15]

More modest buildings, closer to street life, are no less thickly cloaked in wooden screens. A house at 21 Via Matteoti (figure 4.8), the main merchandising street of medieval Volterra and the *via pubblica* named in the 1210–22 statutes, has beam-holes and brackets for galleries at two levels, and doors to the galleries that have now been partially bricked-up to the size of windows. Above the shops at ground level and at the top of the wall, above the galleries, are the fixtures for pent roofs. A building in a more peripheral location, on the street to the town's east gate (Via Don Minzoni 44) has a tall ground storey that may have served as a merchant's *fondaco*, or warehouse, and a gallery on the upper level, to which a doorway gave access and which screens a very handsome double-light window (figure 4.9).

Even later medieval buildings were well-wooded externally. In Volterra, the best example is the palace that, according to a rare inscription at the entrance to the house, the master mason Giroldo di Jacopo da Lugano built for the nobleman Giovanni Toscano in 1250 (figure 4.10). The residence was built around an existing tower sited at an important intersection. The new construction included an entrance hall at the ground level which contains the lower flights of the stairs that continue inside the tower. In the two storeys above ground the addition consists of L-shaped halls wrapped around the tower. The upper hall, as is common in tower and tower house construction, is vaulted. The thirteenth-century structure continues the gallery lines of the tower –as marked by holes and brackets for supporting beams –and there are a number of doorways at the upper levels that gave access to the external

81

Figure 4.8 House at Via Matteoti 21, Volterra (photograph by the author, 1991).

spaces. At the very top of the building, channels that drew smoke from the open fires lit in the hall penetrate the wall in narrow slits well above the level of the galleries and of the pent roofs that sheltered them (Consortini 1942).

In many central Italian towns, the stone skeletons of residential buildings preserve the marks of the galleries, jetties and pent roofs that dominated the medieval streetscape. In no domestic architecture, however, was the balance between stone structure and the secondary elements hung from it weighted so heavily in favour of the jetties as in Pisa. A building type appeared in this early entrepôt between Europe

Figure 4.9 House at Via Don Minzoni 44, Volterra (photograph by the author, 1991).

and the east in the twelfth century to accommodate the houses and storage facilities of traders and bonded warehousemen which crowded the streets near the river, reaching to five or six storeys and projecting aggressively out over the street (figure 4.11). Masonry construction was confined to party walls and, on the street side of the building, to a frame of thin piers joined at their summit by arches. At the intervals of the intermediate storeys, the space between the piers was spanned by beams of stone or wood. Beam-holes on the forward faces of the piers show that the floor structures continued without interruption into a jettied area. The enclosing wall was light, sometimes made of wood and straw,

83

Figure 4.10 Casatorre of Giovanni Toscano, Volterra (photograph by the author, 1991).

and probably surfaced with plaster. Except for the piers of the ground-level loggia and the masonry wall above the terminal arches at the top of the building, it was this surface alone that the building presented to the street. House-warehouses on the Via delle Belle Torri, stripped of their projecting walls and closed with conventional more recent masonry, present only the bones of the original buildings. Other structures, perhaps more specifically for residential use, showed more of their hard masonry core. But even these, like the house the Emperor Henry VI gave permission for property owners on the south side of the river to build in 1197, place many facilities, including bathrooms, stairs, as well as

Figure 4.11 House/warehouse on Via delle Belle Torri, Pisa (photograph by the author, 1991).

galleries, roof eaves and gutters, outside the load-bearing wall (Redi 1983: 274–6; 1989). Later residences, like the house that Mosca da San Gimignano commissioned in 1302, accommodated stairs and bathrooms inside the structural wall, but occasional storeys – in this case the third of five – continued to be jettied forward, sandwiched between storeys that conformed to the property and foundation line (Bonaini 1857: III, 185–9; Redi 1983, figure 3).

Buildings constructed with stone frames and enclosing walls of lighter materials can be found in all the cities of Tuscany. They are especially prominent in Lucca, where residential architecture followed the Pisan evolution through pier and arcade façades in the first half of the fourteenth century to a more continuous wall fabric pierced by multiple-light windows in the second half (Pierotti 1960, 1965), but masonry-frame buildings from the thirteenth century also survive in Florence (Mercato Nuovo), Siena (32–4 Via dei Rossi), San Gimignano (Via San Giovanni 57–9), and Volterra (Via Matteoti 21, figure 4.8). In these cities, frame buildings were outnumbered by continuous wall structures at an earlier date than in Pisa. The Volterra examples demonstrate this and also the fact that the habit of jettied additions was not tied to a particular structural system. They were everywhere.

Under such conditions, the modern viewer rightly questions whether these buildings really faced the city in an architectural sense. Was the new public street, which was potentially such a wonderful theatre for architectural display, wasted on its own inventors? The answer is no; but the ways in which Italian residences presented themselves to the street in the Middle Ages were very different from the way they have done it since the Renaissance.

LATE-MEDIEVAL PALACES AND THE STREET

What the early residences had were a few elements that received special ornamental treatment. In Viterbo the *proferlo*, an external stone stair and its landing, was one of the principal ornaments of domestic architecture. The late-thirteenth-century Palazzo Sacchi preserves an especially monumental example (figure 4.12). It projects 5 feet (1.5 metres) over the street and is supported by columns, arches and substantial stone corbels all decorated, as is the great arch that opens into the ground floor loggia, with finely cut mouldings. Whether or not the upper levels of this building, remodelled in the sixteenth century, were originally screened by wooden structures, the architecture at ground level already distinguished it (Andrews 1982: 48–50 and figure 1.9).[16]

A similar condition holds for contemporary Florentine residences, where the basic form of architectural representation was a prominent display of high-quality construction. Florentine civil architecture is as heavily screened and jettied as any in late-medieval Italy – the projections are called *sporti* here. Twelfth- and thirteenth-century towers, the wing of the Bargello built in 1255 (Paatz 1931: 287ff.), the late-thirteenth-century house of Gherardino de'Cerchi on the Via della Condotta (Preyer 1985) and the slightly later Gianfigliazzi house (figure 4.13) at Santa Trinita, to choose from many possible examples, preserve

Figure 4.12 Palazzo Sacchi, Viterbo (photograph by the author, 1991).

on their forward walls the holes, brackets and metal fixtures for a network of galleries and pent roofs that would have hidden most of the masonry from the street. At ground level, however, behind the work-benches and display tables of the shops, the finely coursed and often rusticated ashlar could be seen.

Not surprisingly, the fancy stonework does not continue up the entire wall. Above the sills of the first full windows the coursing is less consistent and a less expensive surface replaces the finely carved blocks of the rustication. The change is typical. Whatever the surface treatment of the lowest level of a tower in the twelfth or thirteenth century, or a grand house in the thirteenth and fourteenth centuries, the wall above

87

Figure 4.13 Palazzo Gianfigliazzi, Florence (photograph by the author, 1991).

the first window-sill, and sometimes (at the Cerchi house, for example) above the springing-point of the ground-floor arches, is more modestly finished. The fabric may be as rough as simple quarry stone and the surface was most often created with a coat of humble plaster. Plaster was particularly versatile because it could also provide the surface for the thin walls of jetties, as the view of Florence from the 1342 *Misericordia* fresco in the Bigallo (figure 4.14) and the scenes from the life of St Peter in Masaccio's Brancacci chapel show.

The separation by materials of the lower from the upper levels of domestic building was a subject of legislation when the city rebuilt streets and squares at its centre in the late fourteenth century. The

Figure 4.14 View of Florence (part of the *Misericordia* fresco, Bigallo, Florence, 1342).

residences that abutted the enlarged and regularized public spaces had to be reconstructed by their owners in conformity with guidelines established by the city's planners. The regulations concerned only the interface between private property and public space. Repeating legislation first written in 1295 (Moschella 1942: 167–73), the commune took a major step against the zone of privilege it had traditionally accorded property owners by prohibiting all jetties on the renewed public streets. Though enforcement of regulations against jetties was notoriously lax elsewhere in the city, and indeed the tax on *sporti*, far from discouraging them, brought in a revenue of 7,000 florins a year in 1336–7 (Villani 1844-5: III, 321 (book XI, chapter 92)), the commune made the rules stick in these special places. A palace on the Via della Condotta is characteristic in respecting the main street but not the secondary one (figure 4.15). The same is true of a house on the Piazza S. Elizabetta, where a beam-supported jetty survives on the alley but only the scars of a lost counterpart can still be seen on the front of the

Figure 4.15 House at the intersection of Via della Condotta and Via dei Cerchi, Florence (photograph by the author, 1991).

building toward the square (figure 4.16).

The legislation about rebuilding also included specific instructions about walls to be built facing public spaces. In 1363 the owners of *domunculae* on the street around the baptistry of the city 'which degraded the appearance and beauty of the whole square' were forced to remove *sporti* and to build a *muro pulcro* at least 16 *braccia* (9.8 metres) tall on the street side of their property (Gaye 1839: I, 72; Spilner 1987: 387–467). When houses were torn down to expand the Piazza della Signoria toward the north in 1362, owners were required to build 'the wall next to the square beautifully and properly to a height of at least 12

Figure 4.16 House in Piazza S. Elizabetta, Florence (photograph by the author, 1991).

braccia' (7.3 metres) (Frey 1885: 216–17, document 119 from ASF, *Provvisione* 51, c. 81 r., 23 May 1362). Similar regulations were written for the houses bordering the Via Calzaiuoli between the Piazza della Signoria and the Cathedral when it was widened in 1389 (Frey 1885: 240–1, document 140 from ASF, *Provvisione* 79, c. 248 v., 20 October 1389).

The most astonishing thing about these regulations is the modesty of the government's demands. The *faccia dinanzi*, the face or side toward the front, of houses on the most prestigious public spaces in the city were defined only as beautifully built walls. Even first-class construction

Figure 4.17 House at the intersection of Via Calzaiuoli and Via della
Condotta, Florence (photograph by the author, 1991).

did not have to extend as high as the whole building, just 12 or 16
braccia (7.3 or 9.8 metres). Looking at two buildings that fell under the
12-*braccia* ruling we can see just what that meant. On the Via
Calzaiuoli, but not on the side-street fronts of the same buildings, the
wall was constructed as elegantly as it ever was in Florence (figure
4.17). Faced with large ashlar blocks, arched at ground level and
rusticated up to the level of the string course that marks the sill of the
first windows, it exactly fulfils the terms of the legislation. Above that
level, as in the case of the Gianfigliazzi house, rustication was abandon-
ed in favour of less expensive masonry. The residences on the north side
of the Piazza Signoria still display the original fabric of the upper wall
and reveal, in addition to the rustication ending at window-sill level, a
transition from cut stone to stone broken in the quarry at the level of the
springing of the window arches (figure 4.18). Like the *proferlo* of
Viterbo, the decorated structure of the great Florentine houses was
concentrated at ground level. Even though the light-walled *sporti* and
the screens of wooden galleries have been banished from these special
buildings, the habit of treating the upper wall as aesthetically neutral
survives.

92

Figure 4.18 Palazzo Antella where Via Farina enters the north side of the Piazza della Signoria, Florence (photograph by the author, 1991).

THE IDEA OF THE FAÇADE

The modern attitude about the street fronts of urban buildings is very different. Since the Renaissance, the street wall has been understood as a façade. No longer merely part of the structure or even just the front of the building, the façade, as defined by Filippo Baldinucci in the *Vocabulario Toscano* (Baldinucci 1681), had become the element of the building 'which serves the purpose that the face serves in the human body'. It was not only the focus of ornament but also the seat of expression.

The Italian word *facciata*, from which the French and English cognates derive, appears for the first time in the period we have been discussing. It comes from the Latin *facies*, meaning both appearance and countenance or face; yet it is not until the fifteenth century, after *facciata* had come into common use, that *facies* is applied to architecture with the broad meaning of façade.[17] *Facciata* may come more directly from the Italian *faccia* which, in the fourteenth century, had both a geometric meaning, as the face of a polyhedron, and an architectural one (*Dizionario Etimologico Italiano* 1951, s.v.). In 1360, the contract for the construction of the international merchants' tribunal in Florence, the Mercanzia, speaks of the *faccia dinanzi*, the front surface of the building, which was to be finished with carefully worked flat ashlar and decorated with the coats of arms of the city's guilds (Milanesi 1893: 56–7). Arnolfo di Cambio's revetment at the west end of the cathedral, the element we would refer to as the church's façade, is named in 1357 with the same phrase, *faccia dinanzi* (Guasti 1887: 114, document 70, 4 January 1357–8),[18] or simply *faccia* (Guasti 1887: 95, document 70, 21 June 1357).

Facciata enters the language in connection with domestic, not ecclesiastical, architecture. At least, that is the surprising testimony of the evidence I have seen thus far. The first case is Sienese. In February of 1340 the merchant Gontiero Sansedoni signed a contract with three building masters for the construction of a *palazzo* on one of the city's main streets (Toker 1985: 67–95). The document has attracted a great deal of attention because of a unique drawing that accompanies it (figure 4.19). Together, text and image establish the specifications for the commission. The drawing, surprisingly, is not a plan but an elevation and what it represents is the street front of the house, which the text christens *facciata*.[19] The full expression is *facciata dinanzi a strada*, meaning, literally, the façade in front of the street; and despite its cumbersome length, the phrase is repeated *in toto* fully sixteen times. The additional terms are not included to distinguish the street façade from others; the remaining walls abut neighbours or face the internal court and are called, simply, *muri*. It is the newness of the idea, I believe, that explains the fullness of the phrase. Gradually, six times in this document, the short form prevails but for the most part it seems necessary to be complete.

There are other ways in which the concept is still imprecise, or at least different from the one with which we are familiar. For the authors of the Sansedoni contract, the term *facciata* referred to the wall towards the street, in all its aspects. At various points in the text the *facciata* includes structures that are below ground level or out of sight on the inner side of the structure (Toker 1985: 90). Ultimately, of course, visibility, display and design will be central to the idea of façade and certainly they play a role here, but the real focus of the idea in this case is the orientation of the building to the street. There is no more compelling argument for the dependence of the palace façade on the site provided by the new public streets than the phrase *facciata dinanzi a strada*.

The Sansedoni façade design conforms to the same principles of

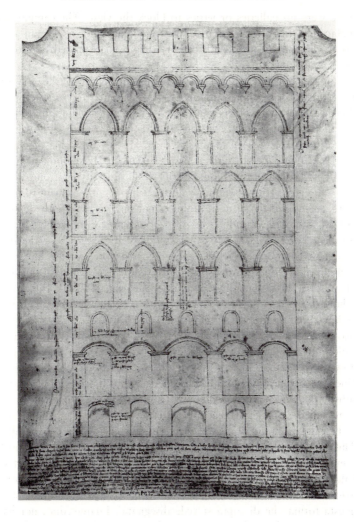

Figure 4.19 Detail of the contract for the Sansedoni Palace (document at Monte dei Paschi di Siena).

ornament that condition the Florentine palaces on the new streets and squares at the centre of the town. The materials in this case are limestone and brick, but the sill of the first full window is still the dividing line and it is the lower level that is built of dressed stone. Gargoyles, the owner's arms, and a corbel table frieze at cornice level ornament the upper surfaces. All of this was on direct display to the street because

95

there were no galleries or jetties to block the view. The campaign to replace galleries with decorative, multiple-light windows, begun at the Campo in the late thirteenth century, has been extended here to the city's main street.

The transformation of the streets of Siena, or of any of the cities that have provided examples in this discussion, into the relatively clear thoroughfares that we experience today was a process that proceeded in fits and starts over many centuries. By the fifteenth century, however, a new direction had been taken. By then, at least for some, it was no longer adequate merely to remove galleries, as the legislation demanded, to adapt a palace to the requirements of 'façade'. At least one Sienese palace owner, petitioning the government for an appointment to public office to support the project, promises not only to remove the galleries but 'to build the façade from the ground up, honourably, like an obedient citizen who desires the honour of the city'.[20] For the more imaginative observer, the change in the streetscape must have been striking. The Emperor Augustus's boast that he had found a Rome built of sun-dried brick and left it in marble (Suetonius 1928: I, 166–7) was echoed by Leone Battista Alberti (1401–72), in the first treatise on architecture and the city since antiquity, the *De Re Aedificatoria*, which was presented to Pope Nicholas V in 1450: 'how many cities, which as children we saw all built of wood [literally, 'crowded with beams'], are now translated into marble' (Alberti 1966: 699 (book VII, chapter 5)). Since, in the early fifteenth century, few central Italian cities preserved substantial numbers of wood houses (Andrews 1982: 2–5),[21] it can only be to galleries and jetties that Alberti refers.

Because he wrote in Latin, the term 'façade' was not available to Alberti, but for Italian-language authors of architectural theory of the same period, *facciata* is fully established. Filarete, writing about the Palazzo Medici in 1462, refers, like the author of the Sansedoni contract before him, to 'la facciata dinanzi verso la Via Larga, la quale sta in questa forma che di sopra si vede disegnata'. Filarete does not have much to say about the Florentine palace, which he must recall from memory as he writes in Milan, nor does he include the promised drawing in his text, but he makes up for it by a more complete account of the Banco Mediceo in Milan, also designed by Michelozzo, the architect of the Medici palace, and executed with great fantasy by Milanese masons. The extended description of this *facciata* (Filarete 1972: II, 698–9 (book 25)), which, in this case, is accompanied by a drawing (figure 4.20), begins by giving the full height and width of the building. In doing this it defines the entire elevation – from street to roof

Figure 4.20 Elevation of the Banco Medicco, Milan (from Filarete, *Trattato* vol. 192 r.).

– as a unified compositional field and goes from there to consider the proportions, architectural vocabulary and ornament by which it is articulated.

The formal rules of façade that lie behind Filarete's description are spelled out in the *De Re Aedificatoria*. It is here that we find the first explicit statement of the primacy of the front of the house. The parts of the building most on display to the public, Alberti writes, deserve the most handsome treatment and anyone who spends so much on the rest of the building (specifically, on its size) that he cannot afford this ornament, makes a fundamental error (Alberti 1966: 783 (book IX, chapter 1)). Contrary to late-medieval usage, Alberti demands that the entire building front be integrated into the design. While there may be more and less intensely ornamented areas of a composition, there are no parts which can be ignored. Decoration should not be 'piled up in single heap', nor should there be any part 'that is neglected and wanting in craftsmanship'. A balance and harmony (*concinnitas*) among all parts represents the ideal (Alberti 1966: 849–51 (book IX, chapter 9)). Even balconies can play a role, as long as they are 'not too large, or ungainly, or too many in number' (Alberti 1966: 811 (book IX, chapter 4)), that is, as long as they are part of the composition.

The residence that Cosimo de'Medici began in 1444 (figure 4.21) has long been recognized as the first great palace of the Renaissance. It is also the first monumental statement of the new attitude toward the façade. The starting-point for the design is the same set of conventions that defined the street fronts of late-medieval palaces. Like the earlier buildings, the ground storey of the Palazzo Medici displays the most

Figure 4.21 Palazzo Medici, Florence (photograph by the author, 1991).

luxurious construction. The difference is that this is rustication at a new scale, so aggressive in its projection and gigantic in the size of the stones that it would appear visually incongruous next to the smooth, thin surface of 'aesthetically neutral' plaster. In the same spirit, its great mass is physically incompatible with the fragile brick or timber-framed wall of a *sporto*. The ground-floor rustication of the Medici Palace demands an upper wall of visual texture and physical substance. As it is, the channelled masonry of the *piano nobile* is only barely strong enough to hold its own against the rustication below. It succeeds because the third level is smoother and 'lighter' still, setting up a progression, more successful between the second and third levels than between the second and the first, that explains the difference. The windows of the Medici Palace, which are the first to introduce decorative tracery to private domestic architecture in Florence, have received a great deal of attention. A pan-Italian perspective[22] makes them seem far less exceptional, but there is no doubt that their detailed carving and the compelling rhythm adds to the visual weight of the upper zones of the façade.

The traditional 'zone of privilege' of the medieval palaces is preserved at the Medici Palace, even if the borrowed space in the street is occupied by very different fixtures. At ground level tables and awnings ceased to have meaning for a palace that conspicuously disavowed craft work and rental income (Goldthwaite 1972; Kent 1987; Cherubini and Fanelli 1990). The arched openings here are for entrances to the palace and a patrician loggia for ceremony and reception. In place of the work surfaces, stone benches occupy the space. The benches had practical and representational functions but, in the context of the history of the façade, what is most important is that, unlike the tables, they were part of the formal composition that extended across the entire street front of the building. At the top of the wall the powerful stone concluding element, the classicizing cornice, differed from the eaves of earlier buildings in a similar way.

The façade that extended the full height of the palace demanded a larger stage on which to project its effects than did the ground-storey ornaments of medieval houses. Public buildings had, for a long time, received visually prominent settings on the city's streets. A street in Parma was enlarged in 1262 'so that the baptistry can be seen' (Braunfels 1953: 127). At the end of the century an approach street to the church of the Umiliati in Siena was widened and the adjacent houses cleared of projecting jetties so that 'the church of those monks and the open door of that church can be seen from the (main public) street'.[23] It is Alberti, again, who theorizes a broader conception of the view. A varied series of prospects, he writes, 'now of the sea, now of mountains, now of lakes, rivers, or springs, now of parched rock or plain, and now of groves and valleys' (Alberti 1966: 665 (book VIII, chapter 1)) is the greatest amenity of the country highway. Art intervenes to regularize the roadbed, to tend the surrounding fields, and, finally, to articulate the traveller's experience. Suitably placed watchtowers are an 'imposing sight' (Alberti 1966: 699 (book VIII, chapter 5)), and tomb monuments in the Roman manner catch the attention and instruct the passer-by. In towns, a curving street offers a similar series of visual experiences. 'At every step the visitor is gradually confronted with the face (*facies*, which also means form or appearance, is the word, not the usual *frons*, meaning front) of new buildings and the breadth of the entrance and *prospectus* (view) of every house is arranged in the middle of the street' (Alberti 1966: 307 (book IV, chapter 5)).

RENAISSANCE PALACES AND STREETS

Great cities, however, were supposed to have straight streets (Alberti 1966: 307 (book IV, chapter 5)). Unfortunately, in the cities that the Renaissance inherited from the Middle Ages, only rarely were they also adequately wide. The solution was to reverse the process of the preceding century when the palace as a building type came to terms with its site along the public street, by adapting, instead, the street to the palaces' new presence. Builders were at pains to establish a setting that allowed the viewer to read the entire façade composition, but they had to do it without abusing the public ownership of the streets. The result was a process of purchase and exchange of land that functioned according to terms established by the public authority.[24]

Filarete tells us that the owners of the Banco Mediceo, having completed the palace, were in the process of widening the street on which it faced. 'And I don't doubt that once the adjacent houses are down the façade will appear more magnificent and much more beautiful' (Filarete 1972: 703 (book 25)). In a similar spirit, the Florentine merchant Giovanni Rucellai took it upon himself to open up the site in front of the palace that he created from a group of family houses by some rebuilding and the construction of Alberti's famous façade in the period c. 1455–8. In the process of building a loggia for family ceremony, he demolished a shop opposite the palace and opened up the intersection of the Via della Vigna Nuova with the narrow Via del Purgatorio. The triangular piazza (1456–63) created a handsome setting for the new façade and opened a view of the building and, especially, its entrance for visitors approaching from the important area around the church of Santa Trinita (Preyer 1977; Kent 1972; Taddei 1973) (figures 4.22 and 4.23). At the end of the century, the Palazzo Strozzi (begun 1489) which is both the biggest of the Florentine Renaissance palaces and the one built in the most crowded and central site in the city, was saved from a setting in which the design of its façades would have remained largely unreadable by the clearing of a square (figure 4.24). The 17.6-metre wide strip of space in front of the palace widened an existing street and lengthened the family *curia* (an open space at the centre of the residential compound of the extended family). Filippo Strozzi, who began the palace, initiated the purchase of property in the area of the new square but it was only his heirs, in the 1530s, who cleared the site. The intimate tie between the square and the palace is reflected in its status as private property. The Strozzi fought into the nineteenth century to maintain their rights to this space (Elam 1985).

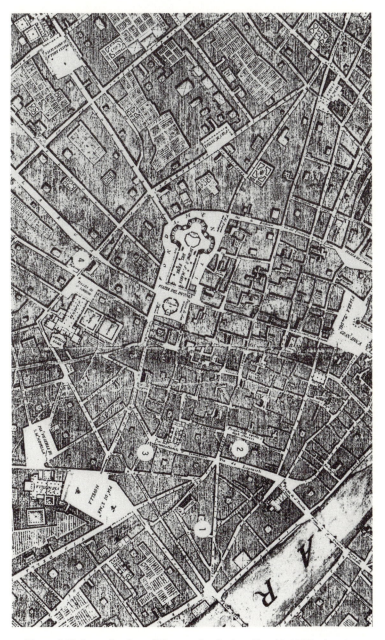

Figure 4.22 Part of a plan of Florence by Cosimo Zocchi, 1783, showing location of palaces discussed in text. 1: Palazzo Rucellai. 2: Palazzo Strozzi. 3: Palazzo Boni (formerly Palazzo Antinori). 4: Palazzo Medici.

Figure 4.23 View of Palazzo Rucellai, Florence, from Via del Purgatorio
(photograph by the author, 1991).

The builders of the Boni (now Antinori) palace negotiated a trade
with the city in 1463 in which they relinquished ground to widen a
street at the rear of the building site in exchange for a triangle of land on
the broad 'boulevard' built over the defensive zone of the Roman town,
the present day Via Tornabuoni, at the front. The new property, still
evident in the plan of the palace (figure 4.25), allowed them to swing
the façade into the street and command a view that extended to Piazza
Santa Trinita almost 400 yards (about 360 metres) away.[25]
    The Medici Palace enjoys a similar prominence on the Via Larga
(now Via Cavour). Significantly, this was not the main street on which

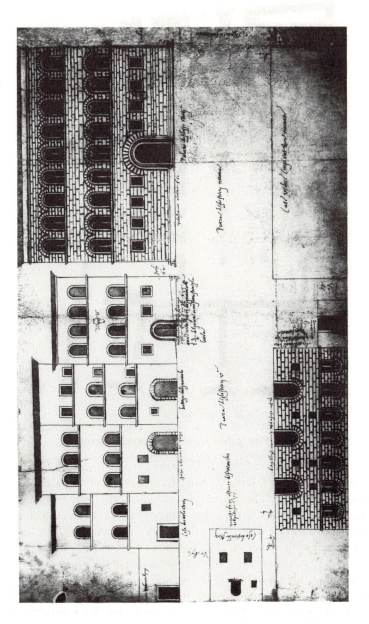

Figure 4.24 Elevation of Piazza Strozzi, Baccio d'Agnolo, Florence, 1533(Ufizzi, Gabinetto dei Disegni e Stampe, 132A).

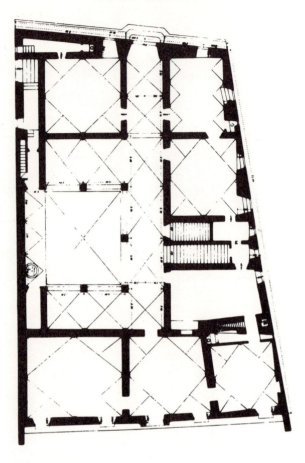

Figure 4.25 Plan of the Palazzo Boni-Antinori, Florence.

the property abutted. It is the Via Sangallo at the back of the palace that ran to the city gate. The Via Larga was built in the early fourteenth century to relieve traffic to the centre of the city and, as its name implies, it was given an ample width. The palace is sited just south of an angle in the west side of the street, an angle that was almost certainly adjusted – perhaps even inaugurated – with the construction of the palace in 1444. The result of this adjustment to the street is that the building seems to turn its face to the visitor arriving from the north. It gains a special prominence at the end of the street and becomes the first landmark on the route to the cathedral, the grain market, or the town hall (figure 4.26).

Figure 4.26 Palazzo Medici, Florence, from Via Larga (now Via Cavour)
(photograph by the author, 1991).

CONCLUSION

In the fifteenth century, private palaces became major monuments in
Florence and other Italian cities, as grand and prominent as most public
buildings. Sited for visual prominence, they began a process that re-
articulated the urban landscape. The revolution was possible only after
a sense of hygiene, social decorum, and physical order, embodied in
town law, exercised pressure on property owners to control the
appendages to their buildings that facilitated private use of the street at
the expense of the quality of public space. The relationship between the

individual building and the town was altered in a fundamental way. Instead of taking advantage of public space for work and habitation, private builders adhered more strictly to property lines and accepted as part of the cost of construction the expenses of an elaborately articulated façade. 'Façade', then, represents a new relationship between buildings and the city. It is a concept born in the fourteenth century and matured in the Renaissance. The Renaissance contributions – the conception of the building front as a unified compositional field and the creation of settings in which façade compositions could have a visual effect – build on the achievement of the late-medieval city where, for the first time, private residential buildings were made to contribute to the decorum of the urban landscape.

ACKNOWLEDGEMENTS

The author gratefully acknowledges the hospitality and support of the American Academy in Rome and of the Massachusetts Institute of Technology that made it possible to prepare this paper. He is responsible for translations into English of quotations from foreign-language works.

NOTES

1   The following abbreviations appear in the text: ASS for Archivio di Stato di Siena; ASF for Archivio di Stato di Firenze.

2   The example of a neighbourhood street, the *contrada del pozo a Sancto Martino* in Siena, is typical both of the kind of problem that involved city officials and of the way they dealt with it. A street 'overcome' by galleries projecting from the houses was to be cleared of the obstructions and regularized if, and only if, the owners, who were also to pay for the work, agreed as a group to the project. The government's street official was placed at the disposal of the abutters to organize the work and could be fined if he failed to execute their decision (Lisini 1903: II, 32 (book III, rubric 48)).

3   The commune of San Gimignano, in its statutes of 1255, appointed public surveyors, called *terminatores terrarum*, who swear 'to decide questions of boundaries of fields, and [building] lots, and the boundaries of houses, and also the division of houses [when divided in wills], and lots, and commonly held ground, and common possessions. We are to make decisions expeditiously and do not have to record them in writing. We are to establish boundaries, and place markers (*termini*) as we see fit, and also to decide questions of private streets not decided by our predecessors, or a judge of the commune, or settled amicably [by the litigants]'. The surveyors were to receive one *denarius* for each marker from the owners of the property

(Pecori 1853: 684–5, from the statutes of San Gimignano compiled in 1255, book I, rubric 48).

4   Lisini 1903: 156–8 (book I, rubric 171). A statute dated May 1292 begins 'for twenty years now the government of Siena has spent large amounts of money to rebuild and repair the streets, bridges and wells in the city and its territory. And because there is no officer to look after these things they are deteriorating and have almost reverted to their original sad state'. The city appointed a *giudice de le vie*, with a staff of three, to enforce the legislation about streets, to collect revenues designated for street work and to oversee repair and the completion of new projects. Citizens were to bring complaints about the condition of streets and roads to the *giudice* who had the authority to force anyone who damaged the roads or usurped them for private purposes to pay for their restoration. All responsibilities for streets and roads previously assigned to the *podestà* were expressly transferred to the new official in this document.

5   A *platea*, or lot, was assigned the new *castellanus* from property in an area of development outside the oldest part of town that was purchased for him by the commune. Even if he bought a property on his own, which apparently some settlers did, perhaps to be able to choose the site themselves, they were still obliged to build within the six-month limit in order to receive the tax exemption. In 1314 an addition was made to this rubric that obliged house builders to spend at least 200 lire on construction.

6   Pecori 1853: 734. Statutes of San Gimignano of 1255, book IV, rubric 74. Prospective builders petitioned the commune for the land and, if the present owner did not commit himself to build on it himself, the petitioner was awarded it for the price of 10 lire per *staioro* with the obligation (and potential fine of 100 *soldi* (about 5 lire)) to complete the construction within ten months.

7   ASS, (see note 1) *Viarii* 1, c. 5 r. (book begun April 1290; the following entry, undated, immediately precedes one of May 1299). 'The *Viarii* command that the ground [*platea*] that lies between the house of the sons of Perino and the house of the deceased butcher Ranucci have its boundaries straightened and its surface paved and that it shall become a public street so that no one may build upon it or project jetties above it and that it shall remain open and unobstructed up to the sky [*usque ad celum*]. The cost of the purchase of the land and of the paving shall be borne by the men of the *contrada* of Sant' Andrea and especially by those who have houses next to the *platea* so that those who get most use and convenience [from the project] shall pay the most.'

8   In early-thirteenth-century Volterra, the space in front of the grand residential structures in the main square was conceded to property owners for permanent, masonry structures serving as benches and steps (Fiumi 1951a: 100–1, from the statutes of 1210–22, rubric 192).

9   Collected from the term *fenestra*, the regulation limited the depth of *deschi*, *banchi*, and *predelle* to a single *passetto*, about 4 feet (1.2 metres). The official had the authority to fine offenders 20 *soldi*. A one-*passetto* limit for wood, and straw stored in the street and for 'things' (*res*) hanging from the building was set by *Statuti* 26 (1337), cc. 249 v. and 250 r., rubric 343, ASS. The *passetto* limit applied to activity as well as to material. Saddle-

makers, barrelmakers, wool-workers, spice sellers and others were specifically enjoined not to occupy more than this space on main streets when they sat in front of their shops to work (Lisini 1903: II, 305–6 (book V, rubric 174)).

10 The awnings and tables for grocers, butchers and fishsellers who displayed their produce on the Campo had to be dismantled every evening. The city charged each vendor 100 *soldi* annually, or more 'if it can be had' (Lisini 1903: II, 31 (book III, rubric 44)). The only exception to the daily clearing of the Campo was the privilege accorded to new knights of the commune, who were allowed to establish a celebratory 'court' on the Campo (presumably amidst the market stalls) for a fifteen-day period (Lisini 1903: II, 35 (book III, rubric 56)).

11 Except in the Campo, sheltering roofs over display tables had to be attached to the shopkeeper's property, whether home or workshop. In 1337 the law became more specific. In the statutes of that year a regulation appeared twice prohibiting mobile display tables in the streets of the city (ASS, *Statuto* 26, c. 250 r. (rubric 343) and c. 253 r. (rubric 370)).

12 In 1309 stairs were excluded only if they restricted access. *Usci*, or doorways which required stairs, were prohibited if they blocked anyone's access to their property or narrowed the public street (Lisini 1903: II, 306 (book V, rubric 175)).

13 Bridges and arches between buildings were exempted from this legislation if they were more than 30 feet (about 9 metres) above ground. A similar law appeared in the 1469 statutes of the less progressive, late-medieval Viterbo. Someone who owned the property on either side of a public or neighbourhood street, or neighbours who mutually agreed to the arrangement, could occupy one-quarter of the width of the street with each of the buildings as long as the construction did not impede horsemen or pedestrians. It was only in a revision of the statute dated 1487 that the incursions into the street were prohibited (Bibliotheca Comunale di Viterbo, II.A. VII. 8 (Statuto di Viterbo 1469 e Riforme), rubric 76, c. 98 v.–99 r. and c. 98 v. in lower margin for the revision).

14 This model study of medieval domestic architecture examines the buildings using archaeological, architectural, documentary and historical evidence. The Via Caprarecce house is discussed on pages 9–12 and a drawing of its façade, with the beam-holes, original doors, windows, niches and stone fixtures clearly indicated, is reproduced as figure 1.7 on page 7.

15 The commune in Volterra regulated tower construction from at least 1207. The legislation included a 10-foot (3-metre) limit to the height of the roofs. This was enough headroom for a belvedere but the roof must also have created a significant impediment to the military operations, especially the use of catapults, for which individuals built tall structures. In fact, the legislation specifically prohibited anyone from fighting from above the roof (Fiumi 1951a: 153–4, Statuto G. I (1224), rubric 92 (dated 1207)).

16 Andrews (1982) uses masonry construction techniques to date the enlargement of the Palazzo Sacchi and the construction of the *proferlo* to the period immediately following the appointment of Giovanni Giacomo Sacchi as treasurer of the Patrimony of St Peter by Boniface VIII (whose family arms appear on the south wall of the building) in 1296.

17   Vitruvius uses the more specifically orientational term *frons* to refer to the porch of a temple (Vitruvius 1970: II, 166 (book III, chapter 2, line 2)). In the 1450s, Alberti uses the same term to refer to the façade of a palace (Alberti 1966: 783 (book IV, chapter 1), and 811 (book IX, chapter 4)), but he also uses the expression *aedificorum facies* to refer to the same element of architecture in a passage where the view of the building is the issue (Alberti 1966: 307 (book IV, chapter 5)).

18   In the same document, which refers to the revetment, or '*pelo*' of the wall, the authors also write of '*faccie dallato*', the side wall, also revetted.

19   There is no reason to believe that this is the first usage for the term, even in Siena, but it could not have been used frequently. When the government prohibited galleries and demanded multi-light windows, after the pattern of the windows of the town hall, on the private buildings surrounding the Campo in the May of 1297 (and repeated the legislation in the statutes of 1309–10), the word *facciata* was not available. Instead the law reads 'qualunque he[di]ficarà casa allato al Campo del mercato, debia fare le finestre a colonnelli et not ballatoio' (Lisini 1903: 116 (book III, rubric 261)). The legislation is repeated in the statutes of 1337, where a change in language gives an index of the progress of the idea of façade, even though the word is still not present: 'si qua domus vel casamentus circa campum fori hedificaretur omnis et singule fenestre talis casamenti et domus que haberent aspectum in campum fori fiant ad columnellas et sine aliquibus ballatoriis fiendis' (ASS, *Statuti* 26, c. 253 v. (rubric 379)).

20   The petition is preserved in the records of the *Ufficiali sopra l'ornato* (1427–80) (ASS, *Concistoro* 2125, petition number 29, granted 25 April 1464; published in Braunfels 1953: 254–5, document 10). In his 1521 edition of Vitruvius, Cesare Cesariano, the Milanese architect and heir to the tradition of Bramante in Milan, translates an ancient concept of building front into a contemporary one and in the process gives a synonym, *frontespizio*, for façade (Bruschi *et al.* 1978: 418). *Frontespizio* was in use at least from the fifteenth century. It appears in a building contract of 1460 for a house in Tuscania in Lazio, where it seems to represent the upper part of a wall that screens the roof gable and shields that side of the house from water running from the roof (Andrews 1982: 94).

21   Andrews (1982) finds little evidence for wooden structures, and a great deal for masonry, in the urban centres of northern Lazio in the Middle Ages. Tuscany had more timber, exporting it through the twelfth century (Herlihy 1958: 25), but must have been running low by the fourteenth century when the trade was banned. Fires reported by the chroniclers of Florentine history in the thirteenth century have often been taken as evidence of wooden construction in the centre of the city. But the amount of wood used for doors, shutters, galleries, jetties, ceilings and floors, roofs, interior walls, and shop furniture in stone structures is sufficient to explain these events. In Siena, galleries had to be made out of brick after 1302 specifically because of the fear of fire (ASS, *Statuti* 26, cc. 178 r. and v. (rubric 315)). Indeed, the framing of load-bearing walls is the least vulnerable of any wooden element in a building. Documents that identify the materials of houses in the late Middle Ages primarily name stone or brick. Some very modest buildings were made of packed earth but even they, in

Siena, had to have a street wall of brick (Lisini 1903: II, 406–7 (book V, rubric 409); repeated ASS, *Statuti* 26 (1337), rubric 215, c. 160 v.).

22 I count, as the immediate predecessors of the Palazzo Medici, buildings without galleries, window boxes or jetties (although sometimes with pent roofs over the ground storey entrances) that address the entire street front as a unified field, articulated in a consistent manner. In the early period this treatment is reserved for very exceptional buildings, seats of the greatest families of the city. In Siena the Palazzo Tolomei, as it was rebuilt 1270–2 (Prunai *et al.* 1971), and in Viterbo the Gatti Palace of the middle of the thirteenth century (Andrews 1982: 40–2) have smooth-faced ashlar surfaces and elaborate tracery in the upper-storey windows. Buildings for similar patrons in the period just preceding the construction of the Medici Palace include the Palazzo Gambacorti in Pisa (about 1380) (Redi 1989: 126) and the Palazzo Vitelleschi in Tarquinia (about 1436–9). In Florence, most ambitious late-fourteenth- and early-fifteenth-century palaces (like the da Uzzano-Capponi Palace on the Via dei Bardi, begun before 1411) have the rusticated base and rough masonry or plastered upper levels already discussed. Some – like the Palazzo Davanzatti of the later four-teenth century or the Ilarione de'Bardi-Canigiani palace of about 1430 (Preyer 1983) – have finished masonry higher on the wall, though never to the roof line. Others, often more modest, were plastered from street to eaves (Thiem and Thiem 1964).

23 ASS, *Viarii*, I, beginning 1290, cc. 4 v. and 5 r., rubric 24. Trachtenberg (1988) discusses the building history of the Palazzo Signoria in Florence (1299–1315) and its relationship to the development of the square that was cleared around it in the course of the fourteenth century. It is not until the fifteenth century that the lessons of the precocious façade design of this public monument were taken up by builders of private residences.

24 Arrangements of this kind in Rome in the later Renaissance are the subject of a chapter in Frommel (1973). The grandest project of this kind is the piazza constructed in front of the Farnese Palace in the 1530s (Spezzaferro and Tuttle 1981). For the formation of public spaces in seventeenth- and eighteenth-century Rome within an institutional and social framework, see Connors (1990). A document connected with the construction of the Strozzi Palace provides a good illustration of the exchange of properties between private owners and the city. The government *Deliberazione* conceded all land belonging to the commune within a rectangular site to Filippo Strozzi and received pieces of Filippo's land that fell outside this site 'pro compensatione' (ASF, *Signori e Collegi, Deliberazioni Duplicati*, 25, ff. 388 r.–v., published in Elam 1985: 128, n. 25).

25 The contents of the Boni document were kindly shared with me for this project by Professor Brenda Preyer who will publish it in a forthcoming book on Florentine fifteenth-century palaces.

REFERENCES

Alberti, L. B. (1966) *De Re Aedificatoria* edited and translated by Orlandi, G., 2 volumes, Milan: Il Polifilo.

Andrews, D. (1982) 'Medieval domestic architecture in northern Lazio', in Andrews, D., Osborne, J. and Whitehouse, D. (eds) *Medieval Lazio: Studies in Architecture, Painting and Ceramics*, Papers in Italian Archaeology, BAR International Series 125, Oxford: British Archaeological Reports.

Baldinucci, F. (1681) *Vocabulario Toscano*, Florence: Santi Franchi.

Bonaini, F. (1857) *Statuti Inediti della Città di Pisa dal XII al XIV secolo*, 3 volumes, Florence: G.P. Vieusseux.

Braunfels, W. (1953) *Mittelalterliche Stadtbaukunst in der Toskana*, Berlin: Mann.

Bruschi, A., Maltese, C., Tafuri, M. and Bonelli, R. (eds) (1978) *Scritti Rinascimentali di Architettura*, Milan: Il Polifilo.

Caggese, R. (ed.) (1910) *Statuti della Repubblica Fiorentina, Statuto del Capitano del Popolo degli anni 1322–1325*, Florence: Tipografia Galileiana.

Caggese, R. (ed.) (1921) *Statuti della Repubblica Fiorentina, Statuto del Podestà dell'anno 1325*, Florence: Tipografia Galileiana.

Cherubini, G. and Fanelli, G. (eds) (1990) *Il Palazzo Medici Riccardi di Firenze*, Florence: Giunti.

Ciampi, I. (1872) *Cronache e Statuti della Città di Viterbo*, Florence: M. Cellini.

Connors, J. (1990) 'Alliance and enmity in Roman Baroque urbanism', *Römisches Jahrbuch für Kunstgeschichte* 25: 207–94.

Consortini, L. (1942) *Le case-torri di Giovanni Toscano in Volterra*, Lucca: Scuola tipografica artigianelli.

Degli Azzi, G. (ed.) (1913–16) *Statuti di Perugia dell'anno MCCCXLII*, Rome: E. Loescher.

Dizionario Etimologico Italiano (1951) Florence: Barbera.

Elam, C. (1985) 'Piazza Strozzi: two drawings by Baccio d'Agnolo and the problems of a private Renaissance square', *I Tatti Studies* 1: 105–35.

Filarete, Antonio Averlino (1972) *Trattato di Architettura*, edited by Finoli, A. M. and Grassi, L., 2 volumes, Milan: Il Polifilo.

Fiumi, E. (ed.) (1951a) *Statuti di Volterra 1210–1222*, Florence.

Fiumi, E. (1951b) 'Topografia e sviluppo urbanistico al sorgere del Comune', *Rassegna Storica Volterrana* 14: 1–28. Reprinted in Pinti, G. (ed.) (1983) *Volterra e San Gimignano nel Medio Evo*, San Gimignano.

Frey, K. (1885) *Die Loggia dei Lanzi zu Florenz*, Berlin: W. Hertz.

Frommel, C. (1973) *Der römische Palastbau der Hochrenaissance*, Tübingen: E. Wasmuth.

Gaye, G. (1839) *Carteggio Inedito d'Artisti dei Secoli XIV, XV, XVI*, Florence: G. Molini.

Goldthwaite, R. A. (1972) 'The Florentine palace as domestic architecture', *American Historical Review*, 77: 977–1012.

Grohmann, A. (1981) *Città e Territorio tra Medioevo ed eta' Moderna, Perugia, secc.XIII–XVI*, 2 volumes, Perugia: Voluminia.

Guasti, C. (1887) *Santa Maria del Fiore*, Florence: M. Ricci.

Herlihy, D. (1958) *Pisa in the Early Renaissance: A Study of Urban Growth*, New Haven: Yale University Press.

Kent, F. W. (1972) 'The Rucellai family and its loggia', *Journal of the Warburg and Courtauld Institutes*, 35: 397–401.

Kent, F. W. (1987) 'Palace, politics, and society in fifteenth century Florence', *I Tatti Studies*, 2: 41–70.

Lisini, A. (1903) *Il Costituto del Comune di Siena Volgarizzato nel MCCCIX–MCCCX*, 2 volumes, Siena.

Milanesi, G. (1901) *Nuovi Documenti per la Storia dell'Arte Toscana*, Florence: G. Dotti.

Moschella, P. (1942) 'Le case a "sporti" in Firenze', *Palladio* 6, 5–6: 167–73.

Paatz, W. (1931) 'Zur Baugeschichte des Palazzo del Podestà in Florenz', *Mitteilungen des Kunsthistorischen Institutes in Florenz* 6: 287–321.

Pampaloni, G. (1973) *Firenze al tempo di Dante: Documenti sull'Urbanistica Fiorentina*, Rome: Ministero dell' Interno.

Pecori, L. (1853) *Storia di S. Gimignano*, Florence: Tipografia Galileiana.

Pierotti, P. (1960) 'Ricerca dei valori originali nell'edilizia civile medievale in Lucca', *Critica d'arte*, 39: 183–98.

Pierotti, P. (1965) *Lucca: Edilizia Urbanistica Medievale*, Milan: Edizioni di Comunità.

Preyer, B. (1977) 'The Rucellai loggia', *Mitteilungen des Kunsthistorischen Institutes in Florenz* 21: 183–98.

Preyer, B. (1983) 'The "chasa overo palagio" of Alberto di Zanobi: a Florentine palace of about 1400 and its later remodelling', *Art Bulletin* 65: 387–401.

Preyer, B. (1985) 'Two Cerchi palaces in Florence', in Morrogh, A. (ed.) *Renaissance Essays in Honor of Craig Hugh Smyth*, Florence: Giunti Barbera.

Prunai, G., Pampaloni, G. and Bemporad, N. (1971) *Il Palazzo Tolomei a Siena*, Florence.

Redi, F. (1983) 'Dalla torre al palazzo: forme abitative signorili e organizzazione dello spazio urbano a Pisa dall' XI al XV secolo', *I Ceti Dirigenti nella Toscana Tardo Comunale*, Comitato di studi sulla sotria dei ceti dirigenti in Toscana, Atti del III Convegno: 5–7 December 1980, Florence: F. Papafava.

Redi, F. (1989) *Edilizia Medievale in Toscana*, Florence: Edifir.

Saalman, H. (1990) 'The transformation of the city in the Renaissance: Florence as model', *Annali di Architettura, Rivisista del Centro Internazionale de Studi di Architettura Andrea Palladio* 2, 73–82.

Salvagemini, G. (1976) 'La Guerra degli sporti', *Granducato*, 1: 3–12.

Sanpaolesi, P. (1939) 'Un progetto di costruzione per una casa del secolo XIV', *Atti del IV Convengo Nazionale di Storia di Architettura*, Milan.

Spezzaferro, L. and Tuttle, R. (1981) 'Place Farnese: urbanisme et politique', *Le Palais Farnese*, Rome, 1, 1: 85–123.

Spilner, P. (1987) ' "Ut Civitas Amplietur". Studies in Florentine urban development 1282–1400', unpublished PhD dissertation, Columbia University, New York.

Suetonius, T. C. (1928) *Suetonius with an English translation*, edited and translated by Rolfe, J. C., 2 volumes, London: Heinemann.

Taddei, D. (1973) 'Piazza Rucellai in Firenze', *Studi e Documenti di Architettura*, 3: 11–48.

Thiem, G. and Thiem, C. (1964) *Toskanische Fassaden-Dekoration in Sgraffito und Fresko, 14. bis 17. Jahrhundert*, Munich: F. Bruckmann.

Toker, F. (1985) 'Gothic architecture by remote control: an illustrated building contract of 1340', *Art Bulletin*, 67, 1: 67–95.

Trachtenberg, M. (1988) 'What Brunelleschi saw: monument and site at the Palazzo Vecchio in Florence', *Journal of the Society of Architectural Historians*, 47, 1: 14–44.

Villani, G. (1846) *Cronica di Giovanni Villani a miglior lezione ridotta coll'aiuto de'testi a penna con note filologiche di I. Moutier e con appendici storico-geografiche compilate da Franc. Gherardi-Dragomanni*, Florence.

Vitruvius (1970) *On Architecture*, edited and translated by F. Granger, 2 volumes, Cambridge, MA: Harvard University Press.

5

MORPHOLOGICAL CHANGE IN LOWER MANHATTAN, NEW YORK, 1893–1920

Deryck W. Holdsworth

INTRODUCTION

The gradual infilling of persistent medieval property parcels in many European cities has made it possible to reconstruct urban growth and economic change by meticulous attention to morphological detail. A property may have been the site of a number of buildings through several centuries, and the amount of building coverage on a plot may ebb and flow, as the repletion cycle for burgages suggests (M. R. G. Conzen 1962). Nevertheless, property parcels (and street outlines) seem eternal. For the most part, this is because constraints of building technology often kept structures within existing property envelopes; the property and land markets were stable; and the pace of economic change was relatively slow (Vance 1971). By contrast, in the downtowns of American cities, the pace of change was rapid; the scale of rebuilding was revolutionary as new technologies developed; and the property market was volatile in an era of adjustments to the demands and opportunities of an expanding national urban system (M. P. Conzen 1987; Vance 1990). Original property parcels were often quickly obliterated. Demolition rather than adaptive reuse has long been the norm in American cities and, as key sites were redeveloped at the end of the nineteenth century, horizontal property expansion went hand in hand with vertical growth. In relative terms, therefore, the cadastral frame in American cities does not survive long enough to be used as a container and recorder of long-term historical urban change. Yet this frame is still extremely useful to portray the extent of short-term transformations and, when used in conjunction with archival records of real-estate transactions and corporate business histories, it can assist in the understanding of complex change.

This chapter combines an interest in the analysis of property parcels with an interest in the dynamics of tall buildings to convey transformations of height alongside the transformations of the cadastre. M. R. G. Conzen's plan analysis of Newcastle upon Tyne emphasized building coverage exclusively, but he noted that 'building repletion proceeds not only horizontally but also vertically by an increase in the number of storeys or floor-space concentration' (Conzen 1962: 400). Incorporation of the vertical dimensions would add an exponential complexity to the already complicated analysis of plan. And yet, in this increasingly visual age, when information graphic designers have popularized three-dimensional graphics (e.g. Bertin 1983; Tufte 1990), the possibilities and potential of dimensionality for morphological work are enormous. The simultaneous portrayal of plot boundaries, building coverage *and* height in one succinct visual expression paves the way for time-series images that can portray the extent of change in the built environment. Typically, these portrayals have been done for micro-scale examples, such as a street segment or a block (e.g. Gad and Holdsworth 1988, Zacharias 1991). Here the possibilities for an entire downtown are explored. Lower Manhattan is used as a case study, drawing on computer-assisted design programs as a frame for portraying the consequences of a dynamic property market and building programme in the early twentieth century when New York became *the* skyscraper city.

Figure 5.1 portrays the changing built environment of Manhattan between 1897 and 1920. There are hundreds of photographs that can convey the same transformation, many from river level and later from the air (King 1926), but the eye tends to ignore, or cannot see, the presence of a multitude of properties. Instead, it is the continuous streetscape and the variegated skyline that catch the viewer's attention. These computer-generated images, by contrast, retain the fundamental separateness of property and building parcels. The images have been fabricated by digitizing the property parcels and the building outlines from 1:960 and 1:1,800 property maps (Bromley 1899, 1907), and then 'extruding' each two-dimensional polygon into a wire-frame, three-dimensional object with a Graphic Design System (GDS) package. Surface-rendering with a lighting function transforms the wire frames into a mass.

The GDS software enables the viewer to rotate around the image, shown here from the south looking up Broadway and Broad Street (but it is equally possible to look up or down Wall Street or south from City Hall Park) and to change the 'angle of regard' from street level up to plan view. The perspectives in figure 5.1 are the result of experimentation to show the greatest amount of relative height (not too low so that

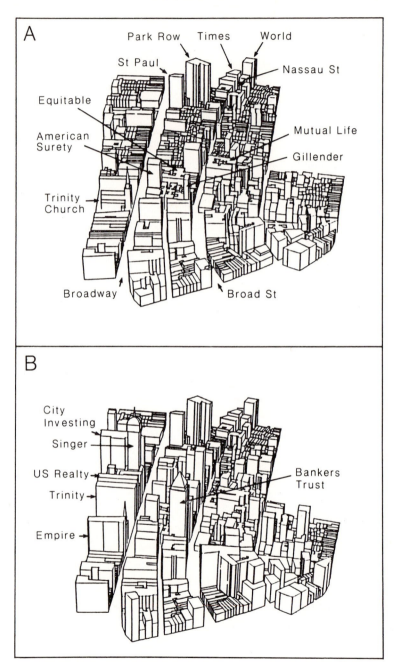

A

Park Row · Times · World

St Paul · Nassau St

Equitable

American Surety

Mutual Life

Gillender

Trinity Church

Broadway · Broad St

B

City Investing

Singer

US Realty

Trinity

Empire

Bankers Trust

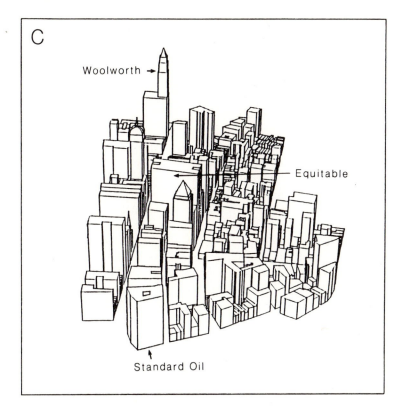

Figure 5.1 Changes in building height and building volume in part of lower Manhattan, New York city. A: 1897. B: 1913. C: 1920.

tall buildings can only just be seen over a four-storey high roofscape, and not so high that the sense of height is lost) and also the most effective vista (with the least number of buildings lost behind taller ones). In order to minimize digitizing fatigue and to simplify the time-consuming calculations necessary to produce each new perspective, it was decided that blocks on the fringes of downtown (west of Trinity to the Hudson River, and east of Nassau to the East River) would not be treated. These blocks, not represented in figure 5.1, remained largely intact with four-storey, nineteenth-century structures on 25- or 30-foot (7.6- or 9.1-metre) frontage lots.

PORTRAYING CHANGE

Although it is the heights of new structures that are first noticed in these computer-assisted images, the specifics of height are hard to fix. From the property maps it is possible to calculate the number of new buildings, and it is useful to know that there were 5 buildings with over 20 storeys in 1897 and 62 between 10 and 20 storeys. By 1914 there were 32 buildings with over 20 storeys and 135 structures between 10 and 20 storeys high (see also Stern, Gilmartin and Massengale 1983). There was some pattern to these redevelopments.

Figure 5.1A shows the late-nineteenth-century city in a relatively undisturbed state. The dominant building configuration was still a 4- or 5-storey row building on a property parcel of approximately 25 by 100 feet (7.6 by 30.5 metres). The spire of Trinity Church is clearly visible on the west side of Broadway, an area relatively undisturbed by change. The east side of Broadway, however, had already seen considerable change, especially at the southern end (north, and especially south, of Wall Street) where few original rectangles are left. Similarly at the top of the image, where Park Row goes off at an angle at the start of the triangular City Hall Park, the tall skyscrapers such as the St Paul, Park Row, Times and World buildings (25, 30, 13 and 11 storeys respectively) signal, among other things, desirable locations for office buildings (Domosh 1989). On either side of Nassau, two low buildings with central-light courts portray the earlier, more horizontal expansion phases of Equitable Life Insurance (between Broadway and Nassau) and Mutual Life (to the east of Nassau).

By 1913 (figure 5.1B) the extent of change was dramatic. On lower Broadway, Trinity Church has almost disappeared, from this angle at least, hemmed in by tall skyscrapers. All up Broadway and especially on the west side, new buildings that are both taller and larger in scale are evident. Notable are the Trinity and US Realty buildings (20 and 21 storeys, of 1903 and 1906) immediately north of Trinity Church, and then north of those on the west side of Broadway the slim, bulbous-topped Singer Tower (at 47 storeys the tallest building in the world in 1906), and the massive City Investing Building. Immediately south of Trinity Church, the 21-storey Empire Building reflects a rebuilding on the same plot boundary as an earlier 5-storey structure; like the Trinity and US Realty buildings, it was able to take advantage of the far longer plot depth, over 250 feet (76.2 metres), to compensate for a somewhat narrow 40-foot (12.2-metre) frontage. On Wall Street, the pyramidal cap to the Bankers Trust Tower (39 storeys, 1912) helps orient the reader to the cluster of new buildings in the financial district around the intersection of Wall and Nassau.

By 1920 (figure 5.1C), the additional changes are prominently high-lighted by the Woolworth Building (at 50 storeys the new tallest building in the world, finished in 1914) on the west side of Broadway fronting City Hall Park, the massive new H-shaped Equitable Building (40 storeys, 1914) on the east side of Broadway just north of the Bankers Trust Tower, and the massive expansion by the Standard Oil Company at the foot of Broadway. Yet to emphasize those three build-ings would be to overlook and underemphasize the massive amount of change on almost every block. There is very little remaining, in these 40 blocks, of the slim 25-foot (7.6-metre) row structures of the nine-teenth-century mercantile city. In their place, millions of square feet of offices had been developed to serve and orchestrate an industrial economy.

MEASURING CHANGE

Having presented these changes through such macrographics, there is the problem of analysis. It is extremely difficult to focus simul-taneously on 10, let alone 40 blocks, and to begin to assess the extent of change. And yet to focus on individual buildings, especially the land-marks, would be to minimize the very essence of lower Manhattan in these years: the incredible increase of office space piled high in a narrow area. This was a period when American industrial enterprises, banks, insurance companies, railways and retail concerns sought a New York office location, in order to be at the heart of the emergent national urban system (Chandler 1977; Jackson 1984). The island of Manhattan was the hinge between the global and national economies, and agents and decision-makers in both spheres increasingly chose to have a New York presence. For some this was just a single office staffed with one or two people, conveniently close to an exchange or a dock, but for others it was a head office with hundreds of employees. Both kinds of demand translated into thousands of offices being built each year, some for rent and others for ownership. As developments in transportation (elevated railways and subways) and building technology (newer, faster elevators; better lighting; telephone systems and so on: Banham 1969) came into being, the marketing of space in new buildings and new locations became highly competitive (Schultz and Simmons 1959).

Considerations of the links between the skyscraper and the social forces in industrial and corporate capitalism that made and remade lower Manhattan are treated elsewhere (Fenske and Holdsworth 1992; Zunz 1990). In this chapter, the focus is more narrowly on morphological detail, namely changes in plan, height and volume. In each case, examples of particular properties, styles or corporations are used to illustrate the broader patterns. Lower Manhattan, it must be noted, does not lend itself to generalizations beyond each block, unlike many other American downtowns with standard-sized, square or rectangular street blocks. The original Dutch walled settlement, with its fort and triangular Parade (later Bowling Green) at the foot of a narrowing Broadway, the former canal or gutte filled in as Broad Street, the irregular width of Wall Street from the line of the old fortifications, and a mixture of alleys, lanes and filled-in waterfront slips make each parcel distinctive (Reps 1965). Lot depth varied and lot widths were hardly standard. Consequently, calculations of average size or average volume for one block do not lend themselves to comparison with other blocks. Similarly, some structures were unique outcomes of specific corporate building histories. None the less, enough blocks contained some mix of generic and distinctive buildings to suggest the pattern of development throughout.

Seen in two dimensions, from a plan perspective, the overwhelming theme is the change in the number of parcels per block. Blocks with more than a dozen lots were reorganized down to a handful of larger parcels, and in some cases consolidation to a single parcel. This transformation proceeded at different paces in different blocks. One block on the west side of Broadway, halfway between Trinity Church and City Hall Park, framed by Broadway, Cortlandt, Church and Liberty streets, was home to seventeen land parcels at the end of the nineteenth century (figure 5.2). Some had small foot frontages (13 feet, approximately 4 metres), most were of 25 feet (7.6 metres), and two parcels along Cortlandt had already been amalgamated (for the Smith Building and the Coal and Iron Exchange). With the exception of the six-storey Exchange, all were at most five storeys high. By 1914, the block was home to only five structures. At the heart of this readjustment was the building programme of Singer Sewing Machine Co., whose head office expansion occurred at the corner of Liberty Street and Broadway. By 1899, their 10-storey office building had been joined by a 14-storey annexe extending west along Liberty Street. As Singer's global empire increased, demands for additional head-office space led to further building. In 1906 the complex was expanded down Liberty Street and north on Broadway. The 14-storey base was capped with a landmark 47-storey tower (Bacon 1986). In plan terms, 11 lots had become one parcel. By contrast, the City Investing Building (1908) was built on three parcels, two themselves the result of earlier amalgamations, and a strategic Broadway lot that gave the building a 'status' address. The massive speculative office building immediately diminished the landmark tower that Singer had just recently completed, boxing in the views from the north- and west-facing windows.

A similar simplification and expansion can be seen on the block east of Broadway defined by Wall, Nassau and Pine where narrow and irregular-shaped lots became far squarer as new tall buildings were developed (figure 5.3). To explore this plan transformation, it will be useful to focus on height, since this was a key issue in the development of the block. The block was home to two distinctive early skyscrapers, the American Surety and the Gillender (visible on figure 5.1A to the right of Trinity Church). At the north-west corner, a 20-storey skyscraper was built by the American Surety Company in 1895 that towered three times higher than surrounding roofs. It was the first building to be designed as a true tower, with architectural detail on all four sides, unlike many others that emphasized the front and almost ignored detailing of the sides on the assumption of neighbouring encroachments. In the case of American Surety, its T-shaped neighbour, the Schermerhorn Building, did threaten to encroach. Its owners objected to the overhanging cornices from the Surety Tower, and threatened to build a 20-storey building that would block the south- and east-facing sides unless the cornices were removed. A compromise led to American Surety's taking over the Schermerhorn, using it for office space, and then rebuilding in the early 1920s on the entire enlarged parcel.

At the other corner of the block, the Gillender Building, at the key corner of Wall Street and Nassau Street, was a 19-storey skyscraper on a mere 25 by 73 foot (7.6 by 22.25 metre) lot. It loomed above the 6-storey Stevens Building that had amalgamated three lots on Wall Street and one on Nassau Street. The Stevens Building became an irresistible location as the financial development of Wall Street intensified, especially when the Bankers Trust Company sought a key site for its head office (Perine 1926). Rather than redevelop the awkward L-shaped plot of the Stevens Building alone, they bought and demolished the Gillender Building, only 13 years old, and developed on a square plot their landmark tower topped by a 7-storey pyramid.

In so far as changes in plan and height are intertwined in these redevelopments, it is appropriate at this point to compare the various ways that such changes can be measured. The Wall–Nassau–Pine–Broadway block was home to ten structures in 1899, at an average

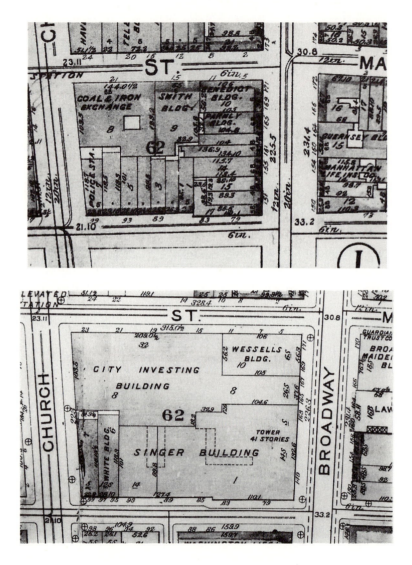

Figure 5.2 Plan transformation, Block 62
(Broadway/Cortlandt/Church/Liberty) (*Bromley Atlas of New York*: New
York Public Library). UPPER: 1902. LOWER: 1920.

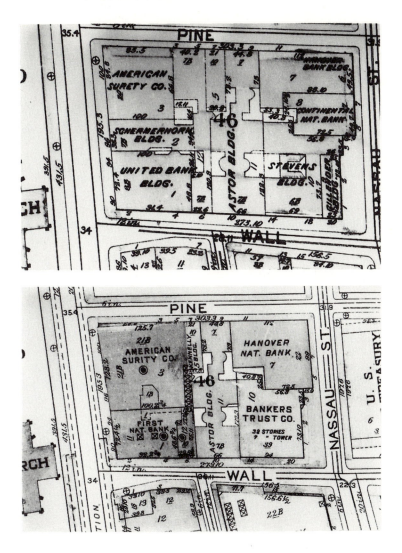

Figure 5.3 Plan transformation, Block 46 (Broadway/Wall/Nassau/Pine)
(*Bromley Atlas of New York*: New York Public Library).
UPPER: 1902. LOWER: 1920.

height of less than 9 storeys (even though the block included a 20- and a 19-storey building); by 1920 just 6 structures occupied the block, now with an average height of 17 storeys. Perhaps it would be more meaningful to say that the amount of floorspace on the block had increased from some 450,000 square feet (41,800 square metres) to just over one million square feet (nearly 93,000 square metres). If this were to be averaged out per building, the change would be from 45,000 to 169,000 square feet (4,180 to 15,700 square metres). Changes on the block can be summarized as a doubling of height but a tripling of volume.

Maps of height alone do not seem to be analytically useful. The number of buildings above 10, or 30, or 50 storeys in any one year certainly is a useful figure for a time series, and these heights mapped do suggest corridors and nodes of important locations. But the Gillender Building at 19 floors had no more floorspace (30,000 square feet or about 2,790 square metres) than its 6-storey neighbour, the Stevens Building, and the tower of the Singer Building was not, in the long run, useful rentable space. Some combination of height and volume may be a more useful way of analysing change. A straight numerical analysis alone of these spaces is certainly unsatisfactory. Consider, for example, statistics for the block on the south side of Wall Street bounded by Broad Street, Exchange Street and William Street. In 1897 there were 17 buildings, with an average height of 6.7 floors and an average space of 26,300 square feet (2,450 square metres), providing 450,000 square feet (41,800 square metres) in total. By 1920 there were only 7 structures providing 907,000 square feet (84,260 square metres) of space (with an average of 113,000 square feet (10,500 square metres), and an average height of 13 storeys, although two were of 25 storeys). Yet such numbers invite comparison with the those for the Wall Street/Nassau Street block (doubling in height, tripling in volume); here it would appear to be a doubling in height and only a doubling in volume. This would seem to be accounted for by the fact that one structure, at the corner of Wall Street and Broad Street diagonally across the street from the Bankers Trust Tower, remained at 4 storeys. This was the head office of J. P. Morgan; in an era of what might be seen as phallocratic expressions of power and importance, his (and everyone else's) knowledge of the enormous wealth of the House of Morgan led to the prestige of a low, 4-storey building for exclusive occupancy (Parnassus 1990). Similarly, further south on the east side of Broadway, the first block of that long street was redeveloped from 14 structures and 800,000 square feet (74,320 square metres) of space to a block with just 6 structures and some 1,355,000 square feet (125,880 square metres).

In this case the transformation did not double volume. One could argue that the way to proceed in the numerical direction would be to find the 'average' change per block for the sample area, then draw a 'location quotient' map to identify which blocks became denser than the norm, less dense, taller, or distinguished by other variables. Such an exercise might be useful if the boundaries remained fixed, but even in this period structures along the Hudson River edge of Manhattan such as the 23-storey West Street Building (1907) and the 32-storey White Hall Building, let alone the later midtown developments such as the Chrysler Building, the Empire State Building and the Rockefeller Center, make it impossible to fix the limits of the office/skyscraper district.

If we return to the Broadway block, it is only when the height and volume changes are calibrated through at least one corporate building programme that the figures make sense. This was the block where J. D. Rockefeller set up a head office following his relocation from Cleveland, Ohio. Number 26 Broadway, home for the Standard Oil Company, became the address for many powerful concerns in the Rockefeller empire. A 9-storey building of 1886 quickly became obsolete and 6 more storeys were added in 1896, as well as a slim 19-storey addition on the lot to the north. Standard Oil still needed additional space, but their expansion was slowed by difficulty in acquiring land on adjacent lots. In 1922 a huge, landmark addition was added to the south, the 24-storey Carrère and Hastings design being constructed in four stages as tenants surrendered their leases (Stern, Gilmartin and Mellins 1987).

UNDERSTANDING CHANGE

This chapter has introduced some aspects of the way that skyscraper office towers transformed the landscape of lower Manhattan. Attention to size and shape of plots, and the height and volume of buildings suggests some avenues for further compilation of information on the texture of this fascinating three-dimensional space. It would be possible to produce a two-dimensional map showing height but, as shown in the case of the Gillender Building, height does not necessarily mean significant volume of rentable or usable space. A map of volume would reflect the speculative agenda of the producers of that space more clearly, but even that requires some detail to give shape to the volume. The conundrum seems to be flat maps, laden with information but not really accessible, or the less usable, but far more emotive, dimensionality of the 3-D image.

As in so many lines of enquiry, the research methods and presentation devices need to be linked with the broader intent. At the heart of an economic and social geography of the changing skyline of Manhattan, I would argue, are the hundreds of tenants that occupy these structures at any one time, and the thousands of people working there as decision-makers, clerical workers, janitors or messengers. Occupancy patterns of skyscrapers are a potent and valuable source for examining a range of economic and social themes associated with the changing skyline. The Singer Tower was the tallest building in the world, but in functional terms the complex was really a 14-storey functional head office with a slim, soon-to-be-seen-as-inefficient, tower of rental office space. Next door was what was labelled the largest office building in the world, City Investing's 850,000 square feet (78,965 square metres), or 'Twelve Acres of Offices' as the *Record and Guide* announced (16 February 1907: 363). On one block, this was the tallest and the largest building of the day. Height and volume were both real-estate advertising vocabularies, along with the variants in façade style, foyer opulence, elevator speed and office configurations, to be used in the speculative office-rental market in the city. The distinction between owning and renting space was confused by the fact that many corporations overbuilt, in order to provide room for later expansion, and were in a position to benefit from revenue-generating tenants in the short term (Gad and Holdsworth 1987). Woolworth took only one floor of the Woolworth Building, Singer executives were hardly in their tower, and workers from American Surety were in only three floors of their named structure. Most of the space was given over to revenue-generating tenants. Some important national firms, such as the Pittsburgh-based Westinghouse Electrical companies, chose not to build their own landmark tower, but instead leased space in the City Investing Building for their export divisions and for key management who worked out of New York. Similarly, various companies in the US Steel empire rented space in the Hudson Terminal Building.

The detailed investigation of the roles of specific skyscrapers as homes for interlocking directorships, or as home to part of this corporate fabric made up of hundreds of lawyers, accountants, advertising agencies or stockbrokers, enables an understanding of the built environment that fuses property, fabric and people. What were the office-space impacts of firms that shifted their headquarters to New York and what were the ripple effects in terms of the supply of and demand for linked services? How footloose were tenants, who typically signed annual leases, as new space became available? The migration of some enterprises took them to six addresses in three decades; how might that be expressed in three-dimensional space?

On the social side, attention to the growing gender and class division of labour in the office sphere (Braverman 1974; Duffy 1980; Zunz 1990) creates further challenges for those researchers interested in morphological analysis. If the size of an office is monitored, in terms of depth away from natural light and air in window bays, or in terms of the location of supervisory men and stenographer women, or the separation of office managers from office workers, then here too the possibilities for imaginative cartography present themselves. Consider, for example, the number of washrooms per floor per building, over time, distinguished by whether they were for men or women; this would be an interesting trace of the historical feminization of clerical work.

The attempt to integrate a consideration of property parcels and a consideration of height has led to the need for flexible, almost snap-shot encounters, giving zoom-lens detail amongst the mass of office towers. The traditional plan analysis does not work for monitoring a variegated skyline of skyscrapers, and the dimensioned cartogram is only impressionistic at the macroscale. Instead a synthetic presentation of material at the scale of the world of a developer, corporation or even tenant seems to fabricate the most profitable strand of meaning with which to weave an effective tapestry.

ACKNOWLEDGEMENTS

I would like to thank the Social Science Research Council's Committee on New York City for funds to purchase copies of the property maps; Matt Tharp, a graduate student in Penn State's Geography Department for his assistance in digitizing the maps and producing the images; and David Shea, of the Architectural Engineering's Computer Design Laboratory, for his assistance in helping us to access GDS.

REFERENCES

Bacon, M. (1986) *Ernest Flagg: Beaux-Arts Architect and Urban Reformer*, Cambridge, MA: MIT Press.

Banham, R. (1969) *The Architecture of the Well-Tempered Environment*, London: Architectural Press.

Bertin, J. (1983) *Semiology of Graphics*, Madison: University of Wisconsin Press.

Black, M. (1976) *Old New York in Early Photographs, 1853–1901*, New York: Dover.

Braverman, H. (1974) *Labor and Monopoly Capitalism: The Degradation of*

Work in the Twentieth Century, New York: Monthly Review Press

Bromley, G. W. and Co. (1899, updated to 1902) *Atlas of the City of New York. Borough of Manhattan*, volumes 1 to 14, G. W. Bromley and Co.

Bromley, G. W. and Co. (1907, updated to 1920) *Atlas of the City of New York, Manhattan, Battery to 14th Street*, G. W. Bromley and Co.

Chandler, A. (1977) *The Visible Hand: The Managerial Revolution in American Business*, Cambridge, MA: MIT Press.

Conzen, M. P. (1987) 'The progress of American urbanism, 1860–1930', in Mitchell, R. D. and Groves, P. A. (eds) *North America: The Historical Geography of a Changing Continent*, Totowa, NJ: Rowman and Littlefield.

Conzen, M. R. G (1962) 'The plan analysis of an English city centre', in Norborg, K. (ed.) *Proceedings of the IGU symposium on urban geography, Lund 1960, Lund Studies in Geography* series B, 24: 383–414; reprinted in Whitehand, J. W. R. (ed.) *The Urban Landscape: Historical Development and Management. Papers by M. R. G. Conzen*, Institute of British Geographers Special Publication 13, London: Academic Press.

Domosh, M. (1989) 'A method for interpreting landscape: a case study of the New York World Building' *Area* 21, 4: 347–55.

Duffy, F. (1980) 'Office buildings and organizational change', in King, A. D. (ed.) *Buildings and Society*, London: Routledge and Kegan Paul.

Fenske, G. and Holdsworth, D. W. (1992) 'Corporate capitalism and the New York skyscraper, 1895–1915', in Ward, D. and Zunz, O. (eds) *The Landscape of Modernity: Essays on New York City, 1900–1940*, New York: Russell Sage.

Gad, G. and Holdsworth, D. W. (1987) 'Corporate capitalism and the emergence of the high-rise office building', *Urban Geography* 8: 212–31.

Gad, G. and Holdsworth, D. W. (1988) 'Streetscape and society: the changing built environment of King Street, Toronto', in Hall, R., Westfall, W. and Macdowell, L. S. (eds) *Patterns of the Past: Interpreting Ontario's History*, Toronto: Dundurn Press.

Jackson, K. T. (1984) 'The capital of capitalism: the New York metropolitan region, 1890–1940', in Sutcliffe, A. R. (ed.) *Metropolis*, London: Mansell.

King (1926) *King's Views of New York*, New York: Manhattan Postcard Company.

Parnassus Foundation/Museum of Fine Arts, Houston (1990) *Money Matters: A Critical Look at Bank Architecture*, New York: McGraw-Hill.

Perine, E. T. B. (1926) *The Story of the Trust Companies*, New York: Putnams.

Reps, J. W. (1965) *The Making of Urban America: A History of City Planning in the United States*, Princeton: Princeton University Press.

Shultz, E. and Simmons, W. (1959) *Offices in the Sky*, New York: Bobbs Merrill.

Stern, R. A. M., Gilmartin, G. and Massengale, J. (1983) *New York 1900: Metropolitan Architecture and Urbanism, 1890–1915*, New York: Rizzoli.

Stern, R. A. M., Gilmartin, G. and Mellins, T. (1987) *New York 1930: Architecture and Urbanism Between the Two World Wars*, New York: Rizzoli.

Tufte, E. R. (1990) *Envisioning Information*, Cheshire, CT: Graphics Press.

Vance, J. E. Jr (1971) 'Land assignment in the pre-capitalist, capitalist, and post-capitalist city', *Economic Geography* 47: 101–20.

Vance, J. E. Jr (1990) *The Continuing City*, Baltimore: Johns Hopkins University Press.

Zacharias, J. (1991) 'The emergence of a "loft" district in Montreal', *Urban History Review/Revue d'histoire urbaine* 19: 226–32.

Zunz, O. (1990) *Making America Corporate, 1870–1920*, Chicago: University of Chicago Press.

Part II

THE NATURE AND MANAGEMENT OF URBAN LANDSCAPE CHANGE

6

SOME ARCHITECTURAL APPROACHES TO URBAN FORM

Micha Bandini

INTRODUCTION

A critical approach to the present complexity of urban morphological issues in British architecture has to contend mainly with two related factors. The first is the ambiguous overlapping of the historical picturesque, vernacular, arts and crafts tradition with an internationally pursued Modern Movement. The second is the lack of any method or concerted approach shown towards the analysis of the physical form of the city by the humanities and social sciences.

This chapter will seek to illustrate, through an examination of proposed and realized design projects and an analysis of investigative approaches to the existing urban fabric, the importance of cultural conventions in shaping both public taste and the way in which research is approached and architectural ideas are transmitted and pursued (Bandini 1984).

Concerns for the relationship between the form of the city, the life of its inhabitants and the appropriateness of strategies for urban planning have, as with any other research subjects, had their moments of prominence as well as abeyance. Often, it is the almost unaccountable surge of 'urban' interest in a discipline which will indicate a change of direction in the character of its discourse.

In the last fifteen years, after a brief flirtation with the interdisciplinarity of the social sciences, and with political issues linked with land ownership or housing distribution, the architectural debate on morphological issues appears to have been channelled in three major directions. First, it was taken up by those architectural critics who, in order to increase their specific competence, shifted gradually from a general commentary upon contemporary buildings into more specific and specialist historical studies. Secondly, it was used by those who saw as paramount the social dimension of architecture and sought to implement it (as in the short-lived 'fashion' of Third World and participation studies), despite the limitations of its scope. Thirdly, it was kept alive and made relevant by those producers of architectural culture who wanted both to give a more solid foundation to their designs and to increase the influence of architecture over society. To this third category belong the international network of architects who, by their teaching, political involvement or cultural concerns, felt that only a theoretical reflection on urban form could ever begin to bring a resolution to the urban problem. These architects were also motivated in their research by two factors: first, the desire to attribute to architecture an autonomous disciplinary status; secondly, the will to give power to socially-motivated architects. Morphological preoccupations, which became part of the international polemic against the Modern Movement, were often the direct result of the personal frustrations of socially conscious architects. Brought up with the utopian ideals promoted by the *Congrès International d'Architecture Moderne* (CIAM), they felt that their disciplinary tools were ineffectual in coping with the complex reality of post-war urban reconstruction, and that the solutions for the contemporary city had been made impossible not only by the economic and political conditions of each country, but also by a lack of debate and investigation.

Different countries reacted differently to what were essentially the same urban issues: the problems of reconstruction, rehabilitation and innovation. Where the planning tradition had been long established, as in the UK, the city became a field to be organized, divided and managed amongst different branches of the planning profession, from highway engineers and planning lawyers to the bureaucratic machinery of central and local government. Where the battle over the city was still largely political, as in Italy, the city remained within the sphere of competence of architects who, often in the disguise of urbanists or politicians, attempted to take a lead in the field. Their education and *Weltanschauung* brought morphological considerations to the forefront of debates. Moreover, it is to the peculiarity of their approach – the way in which they interpreted the historical formation of the city, allowing it to be 'used' for designing – that we owe much of the current international interest in the relationship between urban morphology and building typology.

It would be absurd to expect there to be a shared perspective between the 'theoretically minded' Europeans and the 'pragmatic empiricist' British; nevertheless, it is astonishing to note that the problems faced by each were essentially the same. Even if the breakdown of the relationship between the form of a building and its environment was not perceived in a manner that would now be apparent to us, the international architectural community slowly became aware that the traditional city – the cultural myth of an essentially compact urban tissue that had been able to grow in a more or less cohesive manner through centuries of transformations – had by the 1960s changed in an irrevocable manner.

Nowadays, the ethical and cultural basis of much of the early 1970s research on the relationship between urban morphology and building typology has become almost lost in either a vague notion of contextualism or a plethora of attitudes to design which attempt to verify the existence of, and re-create, traditional typologies. An appreciation that the legitimacy of architecture's form can hardly depend upon historical revival or manipulation has been lacking.

METHODOLOGY

A major focus of this chapter is the examination of the prevalent *mentalité* and its relationship to the perception and production of urban form. For example, the elusive quality of 'Englishness' is invoked in order to explain why, in the United Kingdom, morphology has played such an unfashionable part in the contemporary enquiries of several disciplines, and why, to a large extent, architecture is not at the forefront of cultural production, which may explain why the city and the form of the city, in particular, have not been at the centre of British political and cultural concerns. The main issue is the joint presence in British culture of two intellectual strands; one empirically rooted in traditional preoccupations, habits and forms (of the mind as well as of the environment), the other more rationally based on the international debate, opening up to non-indigenous currents of thought and forms of investigation. However this, as with any other generalized cultural dialectic, if not analysed too closely, can help in both creating a scenario for urban preoccupations in Britain, and in unravelling another facet of a prevalent *mentalité*.

As an intellectual position, empiricism in Britain is possibly more part of the past than the present, but nevertheless its influence is still considerable. By underlying most modes of behaviour in private and public life, it creates 'attitudes' towards cultural production and is a pervasive part of the British way of thinking. It thus corrects some of the most abrasive parts of rationalism, such as the need to establish ideas through public polemics or the Marxist belief that culture and politics are both related to the issue of power. Yet not all commentators have been sensitive to the persistence of the empiricist ideology in architecture; a persistence which certainly owes much both to an individual lack of criticism by architects of their own place in the production of ideas, and to a more generalized reluctance to relate architectural history and criticism to historiographical enquiries.

Notable exceptions can be found in the introductory essays of two magazine issues dedicated to the examination of British architecture (Landau 1984, Wild 1977). It is perhaps not accidental that both issues were published under the editorship of Haig Beck, one of the few editors in English-speaking countries who has demonstrated consistent support for critical enquiries of architecture. Wild (1977) analysed the architecture of the 1970s, and clearly addressed the issue of *mentalité*, when he wrote 'the traditional English distrust of theory, and that particular empiricism that equates the real with the ideal, prevents the profession from moving forward' (Wild 1977: 592). Landau (1984) attempts a comprehensive view of the architectural discourses forming contemporary British architecture, and this is a result of the theoretical awareness pursued in the History and Theory Programme at the Architectural Association Graduate School in London.

All of this has immediate repercussions for the understanding of the form of the city, both in the way in which its problems have been tackled and in the way in which researchers have chosen to represent them. Only by understanding the *mentalités*, the cultural conventions, the social or intellectual consensus in which ideas are received and born, will it be possible to account for the apparent ambivalence towards urban form in Britain. By remembering that an empiricist *mentalité* assumes that a problem-solving process is central to design, its architectural forms being derived from the hierarchy of its brief, we can better understand the full extent of the early division of competence between architects and planners in urban matters. In Britain, this division between disciplines has been accentuated by the formation of two professional institutions, the Royal Institute of British Architects, which was granted a royal charter in 1837, and the Royal Town Planning Institute, which received its charter in 1914. Urban morphology has rarely been able to find a disciplinary space between the planners, who attempt to solve large-scale issues, and the designers, who tend to operate in smaller dimensions.

This lack of a cultural space for urban morphology is also reflected in those studies which, not within the mainstream of research, nor within its most fashionable branches, but rather at the fringes of several disciplines, have developed their own urban morphological debate since the Second World War. Thus the lack of a concerted approach towards the city in the humanities and social sciences reflects the need for both a theory and a non-academic locus of discussion. Both factors are clearly related to British cultural conventions.

THE FORM OF THE CITY IN BRITISH ARCHITECTURAL DEBATES

The 'Picturesque' debate

In discussing English architectural polemic from 1945 to 1965, Reyner Banham perceived the issue of the Picturesque as being at the centre of architectural debates (Banham 1968). Although Banham's use of the term 'picturesque' is ambiguous, his review is useful as an early perception of conflicting currents of thought in British architecture. The ambiguity is no fault of Banham's, but is inherent in the debate itself. In British architectural debates, the term 'picturesque', aside from its use in the eighteenth century, may refer either to an attitude or a composition (but not a style), or be associated with particular stylistic forms; the former was a polemical usage and the latter a derogatory one, often directed against the Picturesque polemic.

This ambiguity, or cultural convention, provided the foundation for two architectural approaches. One was based on the tradition of picturesque literary theory and landscape practice which provided principles of visual organization. The second was in the social domain, and drew its inspiration from socialist political polemic and used the imagery of the arts and crafts vernacular. Having delineated these approaches, Banham elaborates the argument by introducing some of the intricacies of the modernist debate: from ideas that strictly followed classical modernism to the development of a more 'humanized' society, which would take *place* as well as *space* into account. This second approach became more influential in the development of British culture (and in the argument for the persistence of underlying cultural *mentalités*), because it is within this version of the Modern Movement that architecture was, by putting together the best part of the British tradition with the purism of modernism's first principles, to attempt to resolve the awkward relationship between history and the city that the Modern Movement had left unresolved.

It is important to recall that the cultural conventions of tradition and modernism dominating English architectural thought were never mutually exclusive. It will be shown that both of the two most explicit discussions on the city, that of the Picturesque and that of the Smithsons, were to draw from both the English tradition and international modernism.

Tradition and modernism emerged as a central preoccupation for British culture as early as the 1930s and early 1940s, a time when European modernism showed itself to be a powerful cultural force. Early examples of the importance of tradition for the design of the environment are given by Eden (1934) and Sharp (1935, 1936). British visual culture, historically more interested in the question of the combination of space, rather than in the formal harmony of the individual building, felt the abruptness of the modernist form within the urban context. What had been a necessary component in a European modernist polemic against *beaux-arts* rules of composition made far less sense in a cultural climate that, first, lacked a strong classical tradition, and thus did not require to negate it in order to assert itself and, secondly, had established a long tradition of putting together different styles and ways of organizing urban form within the urban landscape, using principles that had never been strongly prescriptive. Thus modernism, which was mainly to be interpreted as a new way in which to organize space, was to become elaborated within the national character and become intertwined with its traditions. It was this concern with tradition that provided the new Picturesque principles for the organization of the environment that underlay a new visual approach to the city.

A major advocacy for the elaboration of this cultural compromise came from the editorial board of the *Architectural Review*, who claimed that architecture and planning were first and foremost visual arts. At this time, the board included J. M. Richards, Obsert Lancaster, H. de C. Hastings and Nikolaus Pevsner. Their polemic had already begun by the mid-1930s, when the then editor contributed an introduction to the *English Tradition* series (Eden 1934, Sharp 1935, 1936). This took on particular relevance in the aftermath of the Second World War. Owing to the massive destruction of the urban fabric and historic buildings, comprehensive guidelines for conservation and reconstruction became a 'live issue' (see, for example, Richards 1941a, b, 1942; Godfrey 1944).

Because their criticism of contemporary planning was mainly focused on the lack of an aesthetic and emotional dimension (qualities greatly admired in the casual structure of the traditional English town and its architecture), it was understandable that the editorial board of the *Architectural Review* should turn to the Picturesque for principles of visual analysis and design. These principles had been identified in eighteenth-century texts such as Sir Uvedale Price's *Essay on the Picturesque* (1794) or Richard Payne Knight's *An Analytical Inquiry into the Principles of Taste* (1805), which provided a theory of 'giving every object the best possible chance to be itself', encouraging 'significant differentiation' amongst the various elements of the urban landscape and the creation of a rich visual environment through the application of principles such as variety, curiosity, intimacy and contrast. (See Hussey 1927 and Watkin 1982 for detailed reviews of the Picturesque tradition prior to the twentieth century.) The content and the polemical use of these early writings appeared in various articles in the *Architectural Review* during the 1940s. Moreover, the proponents of the contemporary form of this approach advocated the Picturesque as a methodological guideline, not a romantic formal style, which would not only be compatible with the modern architectural idiom but also be able to enrich and improve it.

A concise and comprehensive statement of this attitude, and one which reflects opinions prevalent at the time, appeared in the *Architectural Review* in 1944, in an article by H. F. Clark subtitled 'The art of making urban landscape' (Clark 1944). This, as part of a more general advocacy, also criticized planning proposals for not providing a 'picture' of their schemes.

> There is an aspect of this matter which seems to have been inexplicably overlooked by town-planners. It is the fact, obvious to foreigners and historians, that a natural picture-making aptitude exists among us, and has done for centuries. In Picturesque theory . . . a quite unmistakeable national point of view asserted itself.
>
> (Clark 1944: 127)

Clark's reference to foreigners and historians may well refer to Nikolaus Pevsner who, as an editor of the *Architectural Review* and central contributor to this debate, provided historical essays on the Picturesque and a careful analysis of the unique character of his adopted country. These culminated in his Reith Lectures of 1955, and subsequent book, on *The Englishness of English Art* (Pevsner 1955). However, these were not purely scholarly enquiries. Through the crucial operation of attempting to legitimize the English tradition by the creation of a modern metaphysic to underpin its culture, Pevsner was to enter directly into the architectural debate by advocating 'visual planning' as the only appropriate approach to the city, the only one justified by the tradition of 'Englishness' (Pevsner 1946, 1947).

Various aspects of English tradition received analysis, partly owing to their inherent values, but more especially as an element in the *Architectural Review*'s programme of 'visual re-education'. By 1947, the editors acknowledged the existence of the cultural position that had been running through the publication and finally chose to express it openly, and it is worth quoting at length from their editorial.

> It is to architects that mankind will have to look to realize in visible terms a favourable environment for itself. It is thus the architect's role to become the master co-ordinator, through whom the technicians of the statistical sciences and the mechanical facts, as well as the painters and poets, must look to translate their raw material into the stuff of which visible civilization is made. To this end it is not merely desirable, it is of incalculable importance, that the architect, trained in many arts and sciences, should learn from the painter how to see. In which case the student of sight, far from confining himself to certain limited visual routines ... cannot afford to consider any visual experience a waste of spirit that gives him a wider insight into living patterns. So, too, the architectural magazine. Underlying the whole of the *Review*'s apparently diffuse and disjointed articles on landscape and townscape is this master-proposition of the importance of the pursuit of the visual life.
>
> (Richards *et al.* 1947)

It is only on such vague beliefs that the dominance of a visual 'education' and 'analysis' is founded (related studies include Betjeman 1939, Goldfinger 1941 and Hinks 1955).

An early example of the *Review*'s attitude and of the development of a visual 'taste' can be found in the studies by the painter John Piper, whose romantic representations of urban landscapes focus on the different contexts in which colours, textures and materials are presented (see, for example, Piper 1944). Through sketches, which include notations indicating the visual qualities of the scene, he attempts to convey how certain towns and buildings create a 'pleasing effect'. Although this would appear to have had little impact on the general public, it is important to understand these studies, in the context of the *Architectural Review*'s polemic, as a manifestation of a cultural taste and as an early attempt at that far more influential analytical approach known as 'Townscape'.

If 'Picturesque' provided the content for a visual polemic, then 'Townscape' became the banner for this approach, providing it with a more concise method for analysis and design. The use of the term 'townscape' by the *Architectural Review* appears as early as 1947 (Richards *et al.* 1947), although without the polemical force that it later acquired. The term itself indicates the hierarchical importance of a comprehensive visual approach to the 'town landscape' (thus townscape) and, although this word had previously been used, it was through the *Architectural Review*'s polemic that it became defined as a concept.[1]

Two articles are fundamental for the introduction of Townscape. The first, by H. de C. Hastings, the Director of the *Architectural Review*, writing under the pseudonym of 'I. de Wolfe', reinforces the idea that the Picturesque was the principle underpinning the historical and cultural justification behind the *Architectural Review*'s 'visual education' crusade. Thus he wrote that

> the picturesque philosophy has a contemporary message. It exhorts the visual planner – particularly the English visual planner – to occupy himself with the vast field of anonymous design and unacknowledged pattern which still lies entirely outside the terms of reference of official town-planning routine.
>
> (de Wolfe 1947: 355–62)

The implication of this 'radical visual philosophy' for the planner is that 'whatever the elements out of which the visual scene is built, it is on purely visual and not professional grounds that we as radical planners shall admit or spurn them', and these principles 'can be invoked just as surely when the planner is free to work on an empty site in the modern idiom' (de Wolfe 1947: 355–62).

Such a strong advocacy needed a practical example. The second significant article, Gordon Cullen's *Townscape Casebook* (Cullen 1949), is the first introduction to post-war British architecture of a method of appraisal and design for cities. Through evocative sketches, reminiscent of Unwin's 'framed' urban pictures, which were emphasized by suggestive captions and notes, the 'Englishness' of a tradition was both reinforced in its continuity with the past and transformed into a design tool for the needs of the present (figure 6.1). For Cullen, Townscape is the 'art of relationship', and represents the overall experience and conception of the city. Accordingly, it is important 'to take all the elements that go to create the environment: buildings, trees, nature, water, traffic, advertisements and so on, and to weave them together in such a way that drama is released. For a city is a dramatic event in the environment' (Cullen 1961: 9). To experience this is dependent upon vision, because 'it is almost entirely through vision that the environment is apprehended', and the concept of Townscape provides a method of actually

Figure 6.1 An example of Cullen's evocative sketches, from Cullen 1961.

looking at the city; it can become 'a tool with which human imagination can begin to mould the city into a coherent drama' (London Docklands Development Corporation 1982).

Three factors affecting emotional experience are identified: 'serial vision', which is the unfolding of the town's scenery as one moves through and experiences it; 'place', that is, the physical relation of the human body to the environment (concerned with notions such as near/far, enclosure/openness); and 'content', found in the material aspects of the town fabric and perceived through colour, texture, scale, character and so on. According to Cullen, these are all notions that the popular conception of town planning would trade for symmetry, balance, perfection and conformity.

Cullen's publication *Townscape* (Cullen 1961) presented in an expanded form his various analytical and prescriptive notions, including actual projects, most of which had previously appeared in the *Architectural Review*. Gosling and Maitland (1984), having noted that the date of Cullen's publication almost coincided with that of its American counterpart, Kevin Lynch's book *The Image of the City* (Lynch 1960), wrote:

> A painter by talent and romantic by instinct, Cullen's investigations of the desirable qualities of good urban environments differed considerably from the academic analyses of Lynch. Ironically, the personal vision and graphic fluency which Cullen brought to the explanation of his ideas was to some extent a handicap, arousing suspicion in the minds of those for whom a more 'objective' explanation of the urban designer's purpose was necessary. ... Nevertheless, Cullen's method introduced a rather systematic framework for these sometimes elusive qualities, through the idea of 'serial vision'.
>
> (Gosling and Maitland 1984: 48–9)

Yet those are not mere suspicions. As Alberto Ferrari had already pointed out in 1968, the whole of the Picturesque and Townscape enterprise was based on rather questionable methodological ground, where 'not even the concept of empirical observation is completely free from ambiguity: who is the subject and what are the conditions of observation assumed as real data?' (Ferrari 1968: 292). While we should not be surprised that this criticism came from Italy, a cultural quarter with a very different *mentalité*, the ambiguous reception of Cullen in Britain, and the approach that he came to embody, has been for different reasons. As can be seen from the guarded statements of Gosling and

Maitland (1984), or better from what they choose *not* to notice, the empirical convention underlying British attitudes towards culture is still so pervasive that criticism of the lack of theory can still not be found.

Cullen was commissioned for two very prestigious projects concerning future urban physical forms in the UK: he was in charge of Glasgow city-centre rehabilitation scheme, and was also appointed urban designer to the London Docklands Development Corporation (see London Docklands Development Corporation 1982). In Glasgow his visual analysis outlined a lack of enclosure and definition of the 'edges' of each space in the old town, and thus his recommended designs showed a sequence of courtyards and piazzas. In London, the visual line created by ideally joining the Limehouse church with Inigo Jones's Queen's House in Greenwich became the generative design axis that could bring about a coherent renewal of this large area of dockland.

The London Docklands, being the most important of the Department of the Environment's large-scale schemes of the 1980s, shows how resources and interests have shifted from developments concerned primarily with rural land, through New Town schemes to the redevelopment of large decaying areas of inner cities. Together with Liverpool's Merseyside, the other urban scheme that had been labelled as an 'Inner City New Town', London's Docklands should have been a focus for the best of British inventiveness and research on the creation and management of urban environments responsive to the actual conditions of urban life. However, Wilford's analysis of the London Docklands in the mid-1980s is damning, and has been supported by subsequent events.

> The evidence to date is not encouraging. Regrettably developments completed or in construction and projects published replicate the incoherent and diluted characteristics of suburbia. It appears that land owned by the Development Corporation is being parcelled, cleared, serviced and sold to developers with little regard to the form of development proposed. New housing is predominantly of the 'Noddy' variety with the familiar cosmetic devices (e.g. stepped plan relationships, pitched pantiled roofs and porches) thought to give a commercially safe form and appearance.
>
> (Wilford 1984: 12)

Unfortunately, the redevelopment schemes in both central Glasgow and London's Docklands show the extent to which a lack of direction at the political level has allowed the growth of a series of isolated buildings rather than the building of a new urban morphology. The lack of any clear direction is probably due to the ambiguity of the decision-making process at all levels of implementation. The Docklands revitalization should have been promoted and pursued by the public sector in order to result in more than just a skin-deep visual appeal aimed at prospective corporate clients. The Townscape method has not proved to be the best tool for understanding the complexities of the twenty-first-century city, and has not provided a framework from which a new vision for the urban environment could arise.

Even if Cullen's approach has been superseded by events and is not reflected in the present reality of the Docklands, it is important to analyse it as it so clearly reveals the pervasive British conventions. Cullen focused on the existing features of the Docklands: housing along the river's edge, the boundary formed by the ring road, and the enclosed docks with their large expanses of water: these, together with other existing landmarks such as the already mentioned Greenwich Axis, some external views to the city and some internal views of the dock basins, grain silos and Mudchute (an artificial hill formed when dredging the docks), are identified as those physical elements that will continue to determine the area's visual character. These different aspects are abstracted into an analytical plan that becomes the basis for a proposed Townscape structure (figure 6.2).

Again, Cullen states the intention of revealing the 'underlying drama' of the urban setting. Thus the framework embodied in the 'visual struc-ture plan' outlines three major visual and dramatic sequences: one asso-ciated with the housing community (identifying 'nodes' and 'links'), another with the water basins (identifying 'compression' at a central water square, leading into a 'perspective link', and finally opening toward the tip of the island in an 'atmospheric release'), and the third with the 'scenic or tourist route' (a path from the island's entrance to its tip opposite Greenwich, an 'undeclared axis which will epitomize the Isle of Dogs to the outside world'). These three systems overlap at Glen-gall Bridge, a point which is described as the key to the whole concept.

This visual structure would only be evident when experienced by an individual at eye level – all sketches refer to this impression – thus acknowledging the lack of a formal plan capable of working at different hierarchical scales of design. Both the Picturesque and the Townscape approach, by being over-concerned with the humanizing aspects of urban form, have shown a remarkable lack of interest in those conceptual and physical dimensions that are usually associated with the urban scale.

More generally, the visual approach of Townscape has been criticized by Maxwell for its presumption of an objective perception. It abandons

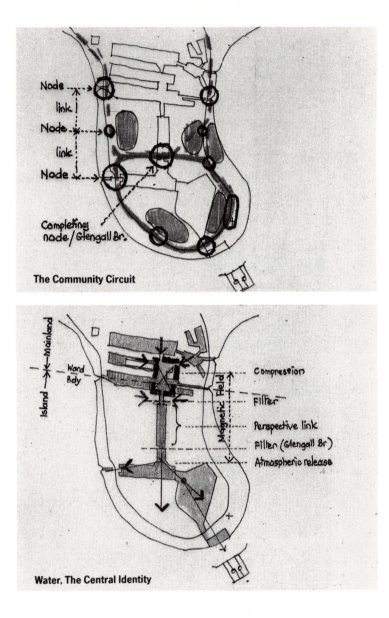

The Community Circuit

Water, The Central Identity

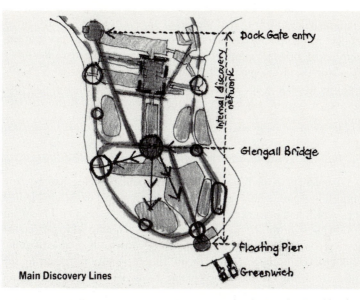

Figure 6.2 Cullen's Townscape-style proposals for the London Docklands, from London Docklands Development Corporation 1982.

cultural associations for visual and emotional effects; a method which, in its requirements for the accumulation of 'visual planning precedents (not principles), by the collection of individual examples of civic design', indicates not only the lack of theoretical structure but the weakness of a method in which discrimination is based solely upon taste (Maxwell 1976).

Aside from its proponents, Townscape has been perceived largely as the manipulation of landscape elements (trees, earth berms, curving roads) and the use of cobbles, awnings, sloped roofs and a variety of materials, colours and textures to improve the urban environment. This reduction of the Picturesque polemic to romantic architectural forms may be the result of an ambiguity inherent both in the use of the term 'picturesque' by its polemicists and in the substance of their contributions. The emphasis on the Picturesque as a principle applicable to modernism is undermined by the *Architectural Review*'s publication of articles that focus on the evocative 'charms' of ruins (for example Piper 1947), or historic and vernacular buildings and towns (Hennell 1941; Donner 1942; *Architectural Review* 1949), which served only to reinforce the view of their detractors that, after all, the *Review* was advocating a preference for historical styles.

147

Notwithstanding Townscape's attempt to provide the distinct terminology and identity that the visual approach lacked, its reception amongst contemporary architects, who saw themselves as being at the forefront of British cultural debates, was rather negative. Pevsner's appeal for the development of a national character through 'Englishness' was not taken up, and for at least two decades, Picturesque and Townscape were anathema in avant-garde circles. For them, the city and its form were to come from a revised version of modernism.

Team 10 and the New Town programme

By the mid-1950s in Britain, architecture and planning had become completely separate fields of competence with regard to the city. Urban form, which had been neglected by both disciplines, became a focus for those who were to see the shortcomings of planning and appreciated the need for a formal architectural language which could, at least at the building level, begin to correct them. The alternative proposed by the *Architectural Review* was unacceptable to anyone who still believed in the social role of modernism and in the development of its ideals, and who saw the Picturesque as the embodiment of those negative values within British culture which the new socialist society was to reject as a basis for planning and design. The Picturesque was, for them, the taste of colonialism, Lutyens's houses, Gertrude Jekyll's herbaceous borders, Cotswold weekend cottages and even, perhaps, croquet on the lawn. All these were upper-middle-class pursuits which a nation that refused Churchill a political mandate after the victory of the Second World War was, for the time being, attempting to erase from the national heritage.

In this context, it is easier to understand the cultural programme of those who saw both the limitations of the 'hard' CIAM line and of its 'soft' British counterpart. Through the rejection of these lines, the new avant-garde were to build their polemic and, by recalling the architecture and neighbourhoods designed by the avant-garde of the 1920s and 1930s, they were to establish, at the same time, the purity of their ideals and a vocabulary of disposable images.

Alison and Peter Smithson were, with Reyner Banham, central figures in the creation of this polemic. They created a set of proposals somewhat less coherent than those upheld by the CIAM elders they criticized, but constituting a programme based on a new 'humanization', and suggesting a concept of 'community', to be built up by various levels of 'association' in order to achieve an 'identity'.

It was this notion of identity that was to become axiomatic in a competition entry that became the banner for a new morphological approach to the city: the Golden Lane project. In this project 'the level of associations' was to become the rationale, that is, man's relationship to house, street and city. But in Golden Lane the existing house and street pattern was seen as no longer being acceptable, since the intensity of motor traffic had precluded pedestrian enjoyment of the street. A new street had to be posited and, for this, a multi-level city with residential 'streets in the air' which recalled Le Corbusier's *Ilot Insalubre 6* of 1939 in Paris and his *Unité d'habitation* of 1948 in Marseilles was proposed (Smithson and Smithson 1967: 22ff.).

The Smithsons' polemic was also concerned with land preservation, which required high-density organization and manifested itself in a multi-level city. Yet while in modernist ideal proposals the same idea resulted in isolated blocks within a geometrically ordered city, and while Le Corbusier's *Unité* was admired in principle, the Smithsons felt that such forms did not provide the human associations that were to be found in traditional urban structures. They also felt that contemporary town-planning theory attempted to create a sense of community by popularizing traditional architectural forms and, by its clustering of functions, was not suited to a modern society that was loosely held together and always socially changing. Thus the Smithsons sought a compromise between the modernist ideals and contemporary town-planning theory:

> the creation of non-arbitrary groupings and effective communications are the primary functions of the planner. The basic group is obviously the family, traditionally the next grouping is the street ... the next the district, and finally the city. It is the job of the architect and planner to make these groupings apparent as tangible realities.

> (Smithson and Smithson 1970: 43)

The Golden Lane project attempted to give these hierarchical units a new urban form. It created family units with gardens located along broad horizontal 'streets-in-the-air', which at vertical circulation and service points would create nodes of activity. At the same time that sense of place and community offered by the traditional city was provided, whilst rising above, and thus preserving, the landscape (Smithson and Smithson 1967: 26–7).

But if the Smithsons' proposals owe much to contemporary socio-logical theory, it is ultimately their 'will to form' that was to create the new morphological patterns. Diagrams of movement and activity

149

patterns, as well as the art of Jackson Pollock and Eduardo Paolozzi, provided the inspiration for the urban form of the vertical housing schemes. These provided 'a random aesthetic reaching-out to town-patterns not based on rectangular geometries, but founded in another visual world' (Smithson and Smithson 1970: 11). The large-scale projects of the Smithsons such as Golden Lane and their small-scale projects such as 'cluster city' followed their 'random' and 'organic' principles. The Golden Lane concept, composed of distinct elements of road systems (straight and direct for easy circulation), ground elements (support services and points of vertical circulation), and the space-defining elements of the dwellings and pedestrian 'streets' (random in pattern, forming important nodes at intersections with other 'streets' or services), is applied to the existing city or landscape, creating 'meaning-ful patterns' by its relation to significant existing urban and natural features. The photomontage of Golden Lane applied to the bombed area of Coventry is especially striking in this respect. By creating new morphological urban hierarchies, what the Smithsons term the 'random aesthetic' responds to the need to orient man within, and establish his relation to, the environment.

The preoccupations of the Smithsons and their Team 10 friends are well documented, but perhaps only in the light of more recent critical contemporary history can the significance of their contribution fully be understood. They were humanizers before being formalizers, but only just! The New Brutalism, both as a polemic and a style (see Banham 1955, 1966 for a discussion of this term), was to become directed at the architectural, but not the urban, debate. Both the ethic and the aesthetic of the Smithsons' Brutalism was to find its way into a series of important buildings and projects, all of which tried to relate the urban to the architectural in a direct and dynamic manner that showed far more awareness than was prevalent at that time.

Influential among the projects spurred by the Smithsons' philosophy was the Park Hill flats in Sheffield (1953–7), designed by Jack Lynn (who had worked for the Smithsons) and Ivor Smith. Also of interest were the new London County Council (LCC) development on the South Bank, the Queen Elizabeth Hall and the Hayward Gallery by Warren Chalk, Ron Herron, John Alleborough and Dennis Crompton (1960–7). These LCC buildings owed much to the Smithsons' thinking on the Berlin Hampstadt competition project of 1958, which was to be elabor-ated once more in the building and pedestrian passages in the Smith-sons' Economist Building Group of 1964.

If, at the architectural level, the Smithsons' criticism had wide repercussions, the same cannot be said for the so-called 'Brutalist connection' advocacy 'to build a city of controlled, pleasurable movement' (Banham 1966). The post-war planning programme still saw New Towns as the main solution for urban redevelopment and expansion; a policy which, while attempting to create new urban configurations, was repeating common housing patterns and traditional relationships between housing and services. The Smithsons' polemic saw only the negative aspects of these arguments. But the role of the New Towns as catalysts for urban research is far more complex than is apparent from the simplification operated by Team 10.

The first-stage New Towns[2] were conceived as 'social environments', unconnected with any rigorous architectural rules but somehow loosely linked with the social architecture of Sweden, which had been evolving while most of Europe had been at war. The 'New Empiricism' (*Architectural Review* 1948) had a social dimension and a formal consistency, but no morphological rules. Thus, in the minds of the New Brutalists, the principles of the New Empiricists, along with the New Towns and Townscape (also being publicized at the same time by the *Architectural Review*), were considered to be of the 'picturesque layout' and 'neighbourhood planning' genre, and thus to be rejected *in toto*.

But the New Empirical influence went far beyond the New Towns. In the LCC, the Roehampton Estate had become the expression of a dispute, with one team of architects (Michael Powell, Clive Barr, Oliver Cox and Rosemary Sjernstedt) producing at one end of the site the Alton Estate East, a romantic, informal group of eleven-storey tower blocks similar to Vallingby, outside Stockholm; while, at the other end of the same site, another team of LCC architects (Colin Lucas, William Howell, John Partridge, Stanley Amis and John Killik) were building their tower blocks and maisonettes in one of the most successful projects ever inspired by Le Corbusier, and the only example of a British approach to modernist urbanism that fully applies the *Ville Radieuse* morphological prescriptions. It is relevant here to note that William Howell, who was part of the Alton West's Le Corbusier-inspired project, was also part of the Smithsons' circle and of Team 10. Such connections can, perhaps, explain the formal antecedents and allegiances of the project.

The later generation of New Towns, beginning in the mid-1950s, was to change many of the architects' and planners' attitudes towards the city. What now needed to be taken into account was the changing nature of transport. This need had already been stressed by the Smithsons in Golden Lane, and Geoffrey Copcutt responded to it in 1955 when designing the centre of Cumbernauld, the only mark-II New Town, as a non-linear structure. This centre was fully serviced by dint of being built over a road, but was also in close proximity to the housing, thus cutting down on what were seen as unnecessary journeys. The entire town was only one mile in width.

The radical critique of the first-generation New Towns, which was to begin with Cumbernauld, was to be strongly reinforced and developed by the Buchanan Report (Ministry of Transport, 1963), and the so-called 'Hook Book', a record of the LCC's attempt at developing a New Town in Hampshire for London overspill population (Greater London Council 1965). Most of the critiques were focused on the then growing concern for flexibility and change, rather than on the need to develop a new formal vision of the city; another example of the inductive-functionalist mode of proceeding that structures the hierarchy of the morphological decision-making process in Britain.

The later, mark-III generation of New Towns[3] clearly showed where the priority of the process was to be. They were conceived not even as discrete formal entities but more as extensions, developments or elaborations of existing settlements which were seen as 'indeterminate', that is, suffering from obsolescence. This was the urban theory that informed the design of the largest and theoretically most sophisticated of the New Towns, Milton Keynes.

The concept behind Milton Keynes, put forward by planning consultants Llewellyn-Davies, Weeks, Forestier-Walker and Bor, was a road gridiron system that divided the designated city area into one-kilometre 'neighbourhood' squares and, in this manner, created a gridded city disposed in relation to three existing small towns and twelve existing villages (figure 6.3). This transport grid system, conceived without an urban dimension (which was expected to come afterwards), was to become, and remain, the main structure in the form of the settlement.

An attempt to accommodate a consistent formal system within the road structure at Milton Keynes was made with different degrees of success in the city centre and in the central area housing. The districts of Fishermead, Conniburrow and Downs Barn achieved some degree of success with this boulevard street architecture; however, the city centre was far more regulated by 'road servicing' and by 'car-parking' than by any morphological constraints of invention. With all of the shopping facilities concentrated into a single building, the city centre was thus to become a built object in a road network, just as the city itself was to become a series of autonomous areas in a macro-road system.

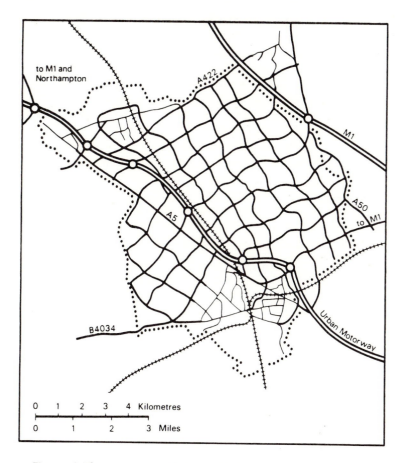

Figure 6.3 The proposed 'gridded city' of Milton Keynes: street plan from Milton Keynes Development Corporation 1970.

There is no doubt that an urban morphological logic was on the agenda of some of the Milton Keynes architects, particularly Trevor Enton and Derek Walker, but the attempt did not succeed. This was not only because of the master plan, which best suited the 'ad hoc' location theory in which there was not even a logical location for the city centre, but also because a compartmentalized plan well suited the then compartmentalized, pluralistic state of architecture, which came to be well represented within the plan (Landau 1984).

More recently, the decline in the public sector enforced by central government has produced almost a standstill in New Town initiatives, and thus active experimentation with urban form is, for the time being, in the hands of the private sector. But, whilst some developers have demonstrated a high level of sensitivity, most pursue the logic of maximum profit. For them, the level of investment required for real urbanism is outside their sphere of operations, and the city is thus now being built piecemeal, in small units that do not have a common plan.

Those architects who were stirred by the modernist versus Picturesque polemic and its development into a debate between New Empiricism and New Brutalism found that there were few ways left open for building the city. A combination of factors was actually preventing them from doing so. Some of the factors were structural, including the economic system of investment and profit in property development; others were superstructural, as in the pervasiveness of a cultural mentality that was able to ignore external influences as long as its fundamental parameters were not threatened: thus

> in England, where in spite of the earlier presence of Marx and Engels, the rigorousness of scientific socialism never succeeded in undermining the reformist tendency of the Labour movement, an aesthetic of cosy domesticity remained the choice of all but a handful, despite the efforts of a 'progressive' establishment. In fact, the 'sentimental' socialism of William Morris, scorned by Engels, has survived to this day, continuing to draw the simple equations of romantic = populist and rational = élitist, with 'people's detailing' as homage to the arts and crafts tradition.
>
> (Wild 1977: 593)

A NEW FORMAL DIMENSION: EUROPEAN PERSPECTIVES

No consistent body of theoretical work on urban morphology can be found in British architecture from the mid-1960s to the present day. Moreover, not only was there no equivalent to the flourishing of the Italian and subsequently the French and Spanish debates, but also European contributions have received little coverage or discussion.

Banham's ambiguous attitude towards 'the Italian retreat from the Modern Movement',[4] because of the way in which it had formed public opinion, had, somehow, precluded a flowering of information. Moreover, while the Italians remained very open to external influences, the British seemed not to take any interest in the new metaphysic that was emerging in Europe.

154

Apart from the well-publicized and discussed sets of proposals for utopian megastructures by Archigram, British urban morphological attitudes are implicit, and can only be read in the light of certain projects and certain completed buildings. In a sense, what is expressed is more important than what is created and, as often occurs in the development of architectural ideas, the unbuilt projects reveal more of the underlying theoretical attitudes than do the built ones.

But there is also a further factor contributing to the lack of British theoretical activity, and of an overt position on urban morphology. That factor is that, until relatively recently, jobs in the practising profession have been readily available for all students graduating from the good teaching institutions. Contrary to the situation in continental Europe, architects who were trained following the Second World War soon had the opportunity to begin building and thus to increase their credibility for both larger commissions and academic appointments. The energy which, in Italy and France, went into historical research, publication and exhibiting – all activities that are conducive to the creation of a theoretical awareness – in Britain most probably went directly into practice, an occupation which, from the nature of architecture itself, is most susceptible to compromise. It is thus natural that teaching had a different flavour in Europe than in Britain, where meta-theories and meta-projects were hardly ever in fashion. Even in the least practically oriented of the British schools, stress was placed upon the inventive necessity of practical requirement blended with formal creation, rather than on the externalization of an *a priori* conceptual construct.

In the Europe of the 1960s and 1970s, three major attitudes characterized the debate on urban morphology. The first used morphology as a means of 'reading' the city. The Venice school and some of the French urbanists are perhaps some of its most conspicuous exponents. It is a useful approach in so far as it clarifies the process of the transformation of cities, and allows urban phenomena to be perceived both diachronically and synchronically. Because it links the morphological nature of the urban fabric with its social, political and demographic aspects, the results are likely to be more comprehensive than any of the products of the disciplines that deal with those aspects individually. But this very comprehensiveness is a by-product, rather than a major objective. The formation of a scientific methodological tool for investigating the relationship between urban morphology and building typology, a recurring theme in the writings of the Italians Rossi and Aymonino, is seen as central to the understanding of contemporary architecture. Its claim to status as a theory that is useful is very weak. 'Understanding' remains the key word, and design is left to another, presumably compatible, line of investigation.

The second approach contains a stronger component of utility: morphology as a way of discussing 'high' architecture in stylistic and cultural terms. Scholarly studies either use architectural type as a tool for comparing or recreating the cultural influences operative at a given time (Wittkower 1952), or treat it as a cultural topic belonging to the aesthetic world, and therefore worth investigating both by itself and in relation to similar categories in painting and sculpture. Treatment of the topic wavers between investigation and interpretation, and it is often only scholarly restraint that prevents typology from becoming a criterion for aesthetic evaluation.

In the third approach, there is a stronger emphasis on morphology as a basis for a theory that is of practical application. Typology becomes a theoretical tool for the production of architecture, and its role is precise. Either it is dealt with in a treatise, as in the eighteenth-century tradition that reached its culmination with Quatremère, or it is the concept that informs a meta-project, as in the case of the work of Rossi, the Krier brothers, and some of the work of Purini, Hall and Unger (see, for example, Steven Hall's 'Typological variations on a rail structure, 1981', published in *Lotus International* 1984).

But inflexible definitions belong to reductionist criticism. In reality, neat distinctions between levels of use, in typology as in other varieties of architectural categorization, are not observed. The desire to employ any notion is such a strong part of the making of architecture, even more so when presented in attractive images, that it overrides other considerations to the extent that a consistently low level of theory[5] is used by most architects to explain their products. What could be more serviceable than a conventionally acceptable theoretical notion such as the relationship between urban morphology and building typology which permits a variety of interpretations with metaphysical overtones?

A useful way to illustrate how the critical is confused with the instrumental in the architectural debate is to follow the development of the relationship between urban morphology and building typology[6] through the different ways in which this relationship is 'used' in order to legitimize architectural projects. I should like to trace the circular and progressive reductionism of 'the usage' that typological research has been made to serve; the argument which, through Argan's interpretation of Quatremère de Quincy's definition of type, has led to the urban studies of Rossi and Scolari; how these have shared a parallel development with Aymonino's notion of type and influenced French urbanists; how typology has been utilized by various 'rationalist' architects; and, finally, how this wealth of theoretical elaboration has been received by the English-speaking culture. It is from this rather convoluted process, over the course of a few years, that the status quo has evolved.

Italian architectural culture of the 1960s was concerned mainly with urban problems. Because their awareness of the limits of the Modern Movement was more pronounced at the theoretical than at the stylistic level, most architects were striving to establish a method capable of uniting the urban with the architectural – a corrective tool for the problems of the city. It is understandable that they should look with sympathy upon Enlightenment architects who, in their break with baroque traditions, established new rules and invented new architectural forms for the changing needs of their societies. It is, therefore, not by chance that the pure forms of Ledoux and Boullée became influential at the same time that Argan presented his article on typology (Argan 1965). A new intellectual mood had been established, focused on the common concerns of a group of architects seeking to discover new practical and intellectual methods on the one hand and, on the other, a historian who looked back to a neglected past. Argan's influence must be considered in the light of the way in which he appropriated Quatremère's definitions of model and type, and offered them as passwords, but also in relation to the more considered reflections prompted by his observations. Designers seem to find in these definitions both the inspiration and the authority for a new architectural 'meta-theory', for a method that reinstates the importance of history and liberates the avant-garde from the immediate past.

Argan demonstrates that typology is not merely a system of classification, but rather a creative process. First, he states that typological series, in the history of architecture, have been formed more for their morphological configurations than for their functional uses. Secondly, he states that the usual means of typological classification is according to the following hierarchical categories: first, the urban scale, with its configurations of buildings; secondly, the building scale, with its large constructed elements; thirdly, the detail scale and its decorative parts. Argan then links these categories with the successive phases of the design of a building, indicating that the uses of typology are not limited to a learned, *a posteriori* evaluation, echoing those hierarchical design stages through which each design is initially conceived and ultimately verified.

One can easily see how this point became influential: it perfectly suited the *modus operandi* of architects. But it is in the argument immediately following that Argan offers more intriguing suggestions, in which he claims that the inherent ambiguity of type, both as an operative tool and as the standard from which one does not depart, makes it applicable to more general questions related to the creative activity versus its history. This, reinforced by his belief that type is always formed through historical experience in its idealized Platonic mode, brings him to the following conclusion.

> The artist, having accepted *a priori* the reduction to type, can free himself from the conditioning influence of a determined historical form, neutralizing it, by assuming that the past is an accomplished historical fact not capable of further development.
>
> (Argan, writing in the *Enciclopedia Universale dell'Arte* (Fondazione Cini, Venezia) 1, 14: 4)

It is not difficult to measure Argan's influence: most Italian urbanists writing between 1965 and 1975 took up his argument, in one way or another. The key article was quoted not only by anyone writing on urban morphology, but also, and more to the point, by architects trying to develop a new meta-theory for design through their drawings. Even his way of using words is echoed in many writings; for example, in those of Rossi and Aymonino, which began to appear at that time. By the time that his article was expanded for publication in the *Enciclopedia Universale dell'Arte*, its authoritative role was undisputed. For the architect of the 1960s who was involved in the painful task of finding an ethically responsible scientific relationship between urban morphology and building typology, Argan's operational typology, combined with Durand's geometrical prescriptions, Quatremère's open definition and Wittkower's analysis of Palladian villas, must have seemed very useful indeed.

In 1970 another book was published that is now fundamental to European urban studies. This was *La Città di Padova*, co-authored by a number of leading Italian architects, whose polemical and left-wing stance impelled them to become involved in urban studies (Aymonino *et al.* 1970). It is here that the main difference between Aymonino's and Rossi's treatment of typology becomes apparent, and the debate engendered by the book begins to acknowledge distinctions between the two methodologies that co-exist within the book itself. While Rossi sees typology as the mediating tool for a formal analysis of the city, Aymonino is more interested in its functional component. He sees typology as an instrument, not as a category. It is understood at two levels: the first is *formal* (independent typology), where it is seen as a means of classification for identifying formal differences, and the second is *functional* (applied typology), in which it is used to understand the endurance of a specific type in the transformation of the city. Aymonino's attitude towards problems of urban analysis, largely unresolved in this first period, becomes more critical in his later work. In *Il Significato delle Città* (Aymonino 1976), a protest against the plethora of drawings derived from the analytical tables of *La Città di Padova*, he clearly condemns the 'naïve results' of those who believed it possible to assemble urban forms from typological analyses. His comment on the nineteenth- and twentieth-century relationship between architecture and urban design – essentially correct, despite its polemical bias – leads him to acknowledge the difficulties built into this method, and this is the origin of his emphasis on the *relationship* between urban morphology and building typology rather than their autonomy. This, perhaps, is the reason why he sees typological analysis not as a low-level theory, but rather as a method, and why he avoids employing the concept as a mechanistic explanation of his projects. Indeed, he states that

> if we start with some contemporary 'deformed' tools of design, such as the coincidence of functional with formal typology, those seen in their architectural stereotypes (kindergarten, hospital, skyscraper, stadium, etc.), I can see that at the urban design stage these methodologies have negated, through their reductive differentiations and the repetitions of their contributions, just that relationship which was most fruitful in urban analysis, just that richness in implications and variety of solutions which, in the end, only architecture can express and solve.
>
> Therefore, for me, analysis and intervention are different tools which find their uses where I try to solve the relationship between architecture and city, between urban form and architectural form.
>
> (Aymonino, in Aymonino *et al.* 1980: 9)

While Aymonino's treatment of typology is still reminiscent of both the Modern Movement and the Italian tradition centred on Quaroni and Samonà, since 1964 Rossi has been writing upon questions related to urban morphology and building typology as part of his attitude towards the design of the city and his rejection of the Modern Movement. Typology was already an important notion in *The Architecture of the City* (Rossi 1966, translated 1982), and even more so in his teaching-note, where his interpretation of type followed Argan's nearly verbatim.

Almost paraphrasing Quatremère, Rossi writes that

> if ... the type is a constant, it can be found in all the areas of
> architecture. It is therefore also a cultural element and as such can
> be sought in the different areas of architecture; typology thus
> becomes broadly the analytical moment of architecture and can be
> characterized even better at the urban level.
>
> (Rossi 1975: 304)

It is at this level that typology, for Rossi, can best be used: it can bridge
the gap between the urban scale and the building scale. In 1970, Rossi
even went so far as to define its usefulness at the political level. 'The
problem is to design new parts of the city [by] choosing typologies able
to challenge the status quo. This could be a perspective for the socialist
city' (Rossi 1970: 9).

A 'correct' political perspective did not come easily to Italian urbanists, divided as they were between the three mainly Anglo-Saxon disciplinary influences and their left-wing allegiances. The political leanings
of Italian urbanists of the recent past are a complex cultural phenomenon. By the beginning of the 1970s, the combined pressure of the
student movement and the economic recession had caused most intellectual architects to question their approach. The alternatives were
either all-out non-architectural political involvement, or a redefinition
of architecture as a scientific discipline. The autonomy of architecture,
one of the canons of the 'Tendenza', was born from the rejection of
1968-style demagoguery and easy urban prescriptions. Typology was at
the centre of this debate. Massimo Scolari's paper of 1971 is particularly
pertinent to this issue. Having quoted Lenin's dictum that 'Truth is
revolution', he proceeded to attack both the consumerism of the bourgeoisie and the pseudo-revolutionary ventilations of the far-left fringes.
He challenged these through a rigorous urban analysis in which the relationship between urban morphology and building typology is the key
factor. Further urban studies have presented the same intransigent line.
Typology, for Scolari, is not a morphological toy. It is a scientific tool
for understanding the dialectics of urban politics (Scolari 1971).

By the end of the 1970s, the methodological tension of urban studies
which investigated typology in a highly charged political and cultural
climate had slackened, and it had become a low-level theory. Nevertheless, a vague consensus about the notion had been achieved in such a
way that its role as an architectural convention, both historical and
contemporaneous, was firmly established. It is interesting to observe
how a cultural convention that had lost its polemical edge in the original

cultural context is exported. As a convention, typology became more powerful through distance and the unfamiliarity of the original language. Original assumptions were no longer questioned, and the distortion of the concepts occurred at the same time that the authority of this badly worked-out, but in some respects attractive, notion was confirmed. Italian urban theories were eagerly adopted by the French, and through them the Swiss and Germans; at first because they responded to a deep cultural need, and later because they soothed generational and political anxieties. For those who had participated, the 1968 students' movement, with its interest in the city, was less reactionary, and thus more acceptable, than an interest in 'object' architecture. The conventional understanding of typology helped to bridge that gap.

On the one hand, it probably appealed to the French critics that their Italian colleagues based their work on Quatremère and Durand. On the other, the relationship between urban morphology and building typology was also supported by a form of alternative history that was rapidly becoming popular in France. The concept of more comprehensive, 'human' history had long been influential, principally through the work of Febvre and Bloch who, through the journal *Annales d'Histoire Économique et Sociale*, formed a school of thought that marginally influenced architects and urbanists.

While this interest in alternative critical methods might have directed research towards minor urban environments and pre-industrial and industrial settlements neglected by the mainstream histories, it is clear that both Castex and Panerai (1979), writing on Versailles, and Devillers and Huet (1981), writing on Le Creusot, explicitly consider Italian urban theories, and Aymonino's applied typology in particular, to be a theoretical source. These two publications exemplify the approach pursued by Bernard Huet in the Institut d'Études et de la Recherches Architecturales. It is through this establishment that most typological research has been disseminated, albeit with an inevitable time-lag: by the time that Castex and Panerai's paper of 1979 was translated into German, typological debates in France had, for the most part, shifted to the line of political disengagement and formal closure already taken by Italian urbanists.

The French situation is instructive. It embraces both politically conscious urban research and formal experiments which, although loosely based upon such research, reflect a completely different structure of beliefs. Panerai's own work exemplifies the dilemma of those who, while inspired by the early Italian urban theories, are nevertheless inclined to search for a functional justification for their design work. Whilst Aymonino's theories have not been completely misunderstood by French writers, it is evident that a great deal of interpretation has taken place. The gap between analysis and project, so clearly acknowledged in the early Italian work, becomes more and more narrow. Moreover, the breadth of the previous work has disappeared, and with it the possibility of verifying typological studies according to criteria external to architecture.

Paradoxically, the very theoretical positions that demanded an architecture related to other types of cultural production are those that forced it to restrict its boundaries in order to define its scope. The issues underlying the Rational Architecture movement are illustrated by work produced and published under the title 'Archive d'Architecture Moderne' (AAM). Maurice Culot and the Krier brothers, perhaps the most articulate exponents of the 'reconstruction of European cities' enterprise, have repeatedly attacked contemporary architecture, its mode of production, and the intellectual and economic assumptions upon which it is based. Type and typology are very much at the centre of AAM publications but, while Robert Krier takes for granted the existence of such concepts and begins his book *Urban Space* (Krier 1975) by establishing a typology of urban spaces, the papers presented in *Rational Architecture Rationelle* (Krier 1978) are still, for the most part, focused on the legitimacy and usage of the notion.

There is, of course, scope for variation within the established boundaries. Certainly Leon Krier's poetic urban inventions reflect a much more subtle imagination than that of his brother's reductivist urban spaces composed of squares, circles and triangles. In the same way Delevoy's introductory essay in *Rational Architecture Rationelle* (Delevoy 1978), in which he calls for an 'operational typology', is much more convincing than Vidler's justification of a 'third typology' (Vidler 1978).

A key factor in this argument is Vidler's editorial in the journal *Oppositions* (Vidler 1968), which focused on rational design processes and, obviously, typology. This was published at a time when the Institute for Architecture and Urban Studies was promoting the work of Aldo Rossi and La Tendenza on the East Coast of the United States. This issue of the journal also contains a revealing series of letters, and a review of a forum held at the Institute, on Rossi's work, which clearly indicate that, at this stage, the East Coast intelligentsia were far from unanimous in their acceptance of European fashions.

Schools of architecture are often the best places to measure the degree of consensus that an idea receives. The progressive reductionism of the typological debate is strikingly apparent in the issue of the American *Journal of Architectural Education* devoted to typology in design education (Morris and Levin 1982). While the editorial begins with the defensive statement that the issue 'does not deal with a subject which all would recognize as essential in design education' (Prologue, in Morris and Levin 1982), it is clear from the list of contributors that typology as a 'vague form of convention' has a place in the most respectable American teaching institutions. The nature of this place, however, is questionable. Absent are the references to the original Italian texts, not then available in full translation. The interest appears to have been focused on type as historical precedent, and current design research completely overlooks the relationship between building typology and urban morphology, regarding typology as only a convenient repository of authoritative imagery waiting to be transformed by personal creativity. As far as typology itself goes, the Rational Architecture movement is partly responsible for encouraging this trend among the English-speaking public. Leon Krier's work and teaching at the Architectural Association in London, and the dissemination of typological notions through *Oppositions*, have resulted in their application in a much narrower enterprise than, perhaps, the authors originally envisaged.

The reductionist process thus far outlined has been most pronounced in Britain and America, where the contributors make only ephemeral appearances and where there are no discourses that can sustain a broad and productive debate. The pragmatic and empirical cultural climate in which we live appears to favour studies that regard typology as a collection of readily adapted icons. The lack of substance and cultural depth in current architectural debates owes much not only to the incompatibility of European rationalism and Anglo-Saxon empiricism (a much-abused generalization) but, as Jorge Silvetti has noted in a survey of the debate,

> it is also because the idea of type in this Anglo-Saxon interpretation does not correspond to that of the tradition of the Enlightenment.... Here we have a purely iconographic interpretation and use of the idea of type. From this perspective, type is far from being an abstraction or a rational principle. Rather it is the cultural icon that appears and circulates in society that is made identifiable and becomes, in turn, the 'represented symbol'.
>
> (Silvetti 1980: 25)

If these are the prevailing concerns, it is understandable that European

163

architectural products have been favourably received in certain US circles that emphasize creative individuality and are not preoccupied with the relationship between urban morphology and building typology.

THE IMPLICATIONS FOR BRITISH URBAN LANDSCAPES

Recent urban research in the UK has developed along very different lines from that elsewhere in Europe. Large-scale issues were tackled at the beginning of the twentieth century, and methods and approaches were largely established by the post-war reconstruction plans. Few contemporary British architects or urban scholars have conceived of the need or the possibility of proposing either a prescriptive theory of the city or a comprehensive analysis of its complexity. The present situation, with its lack of proposals and vitality, has become almost the opposite of the cultural climate which, without interruption from the end of the nineteenth century to the immediate post-war years, was one of creativity and experimentation, involving most of the urban and environmental ideas later to be debated and furthered by other countries. What has been most notably absent is any framework for planning proposals that would further all aspects of urban development. Whilst Berlin, Paris, Frankfurt and Barcelona are busy at the political and economic levels in drawing up strategies that include urban regeneration as a primary objective for financial and cultural supremacy, London has even dismembered its only body for administrative co-ordination. The demise of the Greater London Council is symptomatic of that lack of vision which results in either a populist appeal to an ill-informed public, or to concessions to private interests which pay little heed to real issues in economic and social planning.

In this respect, the debate which has arisen from the stance taken by HRH the Prince of Wales and Paul Reichman, the head of Olympia and York (developers of Canary Wharf), transcends the limits of the chronicle and *de facto* enters into the realm of cultural history.

The capacity to influence has been, and conceivably always will be, formed by the struggle between ideology and power. The urban realm is a direct reflection of this but, whilst in the recent modern past, Britain took the lead to ensure that academic research, political will, technical competence and economic viability were at the service of a vision which, in its comprehensiveness, tended to serve the nation rather than some of its factions, now that will seems lost.

Absent, therefore, from the British approach have been those strong a priori statements that demand design propositions to voice the theories they espouse. Also absent has been that fertile mixture of cultural and political polemic that looks to the city for both its focus of interest and its point of application. Finally, and most importantly, the modernist and post-modern intellectual tradition that finds the stimulation necessary for cultural production only in the metropolis is also absent.

The Anglo-Saxon tradition calls instead for empirically based descriptive studies, carefully worked-out, mathematically based modelling techniques and perceptual analyses that could, by inference, influence the design of future environments. It is a tradition strongly influenced by scientist rather than ideological paradigms, where social consensus is often more the product of accepted taste than of polemical debate, and in which the environment is largely seen as the outcome of a tradition to be complemented but never to be overcome.

The evidence presented in this chapter indicates that this *mentalité*, now stronger than ever, has not found within itself the regenerative power that would bring, even within accepted taste and conventions, both a resurgence of urban interest and the financial, managerial and architectural skills necessary for its implementation. Even if the middle classes have chosen to indulge in a somewhat mystified and sanitized 'Englishness', it is through the establishment's shortsightedness on the historical perspective that Britain is losing the capacity to think about its urban heritage in the same way that its creators regarded it: as a resource and as a symbol of the representation of the present as well as a projection of the past into the future. It is the form of the city that expresses all this through the capacity of its architecture to transcend its own scale and function and to merge into another ensemble. Thus a city form necessitates both the evolving of traditionally accepted conventions and the insertion of new objects – buildings. Urban morphology is ultimately both the study and the creation of such architecture.

ACKNOWLEDGEMENTS

The editors would like to thank Karl Kropf for his helpful comments on a draft of this chapter. Professor Pierre Merlin (Laboratoire Théorie des Mutations Urbaines, Université de Paris VIII) and the Architectural Association kindly allowed reproduction of material from which this chapter was developed. Figure 6.1 is from Cullen, G. (1961) *Townscape*, by permission of the Architectural Press, London; figure 6.2 is reproduced from London Docklands Development Corporation (1982)

Guide to Design and Development Opportunities, by permission of the London Docklands Development Corporation; and figure 6.3 is redrawn from Milton Keynes Development Corporation (1970) *Plan for Milton Keynes*, by permission.

NOTES

1   *The New Oxford Dictionary* cites the previous use of 'townscape' in 1880 and 1889. It should be noted that the term has several connotations: in architectural debates it is an analytical and prescriptive concept, and in urban geography it is a general term of reference. For the purposes of this chapter, 'Townscape' refers to the architectural usage, and 'townscape' to the geographical usage.

2   The first-stage New Towns were Stevenage (1946), Crawley (1947), Hemel Hempstead (1947), Harlow (1947), Hatfield (1948), Basildon (1949), Bracknell (1949) in the London area; Newton Aycliffe (1947), Peterlee (1948) and Corby (1950) outside London; Cwmbran (1949) in Wales, and East Kilbride (1947) and Glenrothes (1948) in Scotland.

3   The mark III New Towns were Central Lancashire (1967), Dawley–Wellington–Oakengates (Telford) (1966), Northampton (1966), Ipswich (1966), Peterborough (1966), Warrington–Risley (1967), Milton Keynes (1967), Skelmersdale (1961), Runcorn (1964), Redditch (1964) and Washington (1964).

4   This is how Reyner Banham characterized Italian post-war building. In a paper published in 1959, he condemned those architects who believed that both history, a new sense of materialism and a novel approach to the historical avant-garde heritage could constitute the way ahead for the post-Second World War generation (Banham 1959). This paper was preceded by another by Portoghesi (1958) and answered by an editorial by Rogers (1959) which began the debate on the relationship between social programmes and formal representation, and on the legitimacy of formal sources, which has recently come to a head in the post-modernist phenomenon.

5   The term 'low-level theory' is here used in the scientific sense, as meaning a theory that is weak on explanatory power.

6   Hereafter, the word 'typology' is generally used instead of 'the relationship between urban morphology and building typology'.

REFERENCES

Architectural Review (1948) 'The New Empiricism', *Architectural Review* 103, 1: 172–4.

Architectural Review (1949) 'The functional tradition', *Architectural Review* 106: 19–28.

Argan, G. C. (1965) 'Sul concetto di tipologia architettonica', *Progetto e Destino*, Milan: 'Il Saggiatore' Alberto Mondadori.

Aymonino, C. (1976) *Il Significato delle Città*, Bari: Laterza.
Aymonino, C. (1980) Untitled contribution, pp. 9–37 in Aymonino, C., Gregotti, V., Pastor, V., Polesello, G., Rossi, A., Semerani, L. and Valle, G. *Progetto Realizzato*, Venice: Polis Progetti, Marsilio Editori.
Aymonino, C., Brusatin, M., Fabbri, G., Lena, M., Lovero, P., Lucianetti, S. and Rossi, A. (1970) *La Città di Padova*, Rome.
Bandini, M. (1984) 'Typology as a form of convention', *AA Files* 6 (May): 73–82.
Banham, R. (1955) 'The New Brutalism', *Architectural Review* 118, 12: 354–61.
Banham, R. (1959) 'Neo-Liberty, the Italian retreat from Modern architecture', *Architectural Review* 125: 747.
Banham, R. (1966) *The New Brutalism: Ethic or Aesthetic*, London: Architectural Press.
Banham, R. (1968) 'The revenge of the Picturesque: English architectural polemics, 1945–1965', in Summerson, J. (ed.) *Concerning Architecture: Essays on Architectural Writers and Writing Presented to Nikolaus Pevsner*, London: Allen Lane.
Betjeman, J. (1939) 'The seeing eye, or how to like everything', *Architectural Review* 86, 10: 201–4.
Castex, J. and Panerai, P. (1979) 'Space as representation and space as practice – a reading of the City of Versailles', *Lotus International* 24: 84–94.
Choay, F. and Merlin, P. (1986) *Á Propos de la morphologie urbaine*, Noisy-le-Grand: Université de Paris VIII.
Clark, H. F. (1944) 'Exterior furnishing or Sharawaggi: the art of making urban landscape', *Architectural Review* 95, 1: 125–9.
Cullen, G. (1949) 'Townscape casebook', *Architectural Review* 106, 12: 363–74.
Cullen, G. (1961) *Townscape*, London: Architectural Press.
Delevoy, R. L. (1978) 'Towards an architecture', in *Rational Architecture Rationelle*, Brussels: Archive d'Architecture Moderne.
Devillers, C. and Huet, B. (1981) *Le Creusot, naissance et développement d'une ville industrielle, 1882–1914*, Seyssel: Éditions du Champ Vallon.
Donner, P. F. R. (1942) 'Treasure hunt', *Architectural Review* 92: 19–21, 49–51, 75–6, 97–9, 125–6, 151–3.
Eden, W. A. (1934) 'The English tradition in the countryside', *Architectural Review* 77: 87–94, 142–6, 151–2, 193–202.
Ferrari, A. (1968) 'The aesthetics of physical environment in contemporary English architectural culture', *Zodiac* 18: 289–95.
Godfrey, W. H. (1944) *Our Building Inheritance: Are We to Use or Lose It?*, London: Faber.
Goldfinger, E. (1941) 'The sensation of space', *Architectural Review* 90, 11: 129–31.
Gosling, D. and Maitland, B. (1984) *Concepts of Urban Design*, London: Academy Editions.
Greater London Council (1965) *The Planning of a New Town*, London: Greater London Council.
Hennell, T. (1941) 'Country craftsmen', *Architectural Review* 89: 53, 85, 111, 132.

Hinks, R. (1955) 'Peepshow and roving eye', *Architectural Review* 118, 9: 161–4.

Hussey, C. (1927) *The Picturesque: Studies in a Point of View*, London: Putman.

Knight, R. P. (1805) *An Analytical Inquiry into the Principles of Taste*, London.

Krier, R. (1975) *Urban Space: Theory and Practice*, Brussels: Archive d'Architecture Moderne.

Krier, R. (1978) 'Reconstruction of urban spaces' in *Rational Architecture Rationelle*, Brussels: Archive d'Architecture Moderne.

Landau, R. (1984) 'British architecture. The culture of architecture: a historiography of the current discourse', *UIA – International Architect* 5: 6–9.

Landau, R. (1986) 'Milton Keynes: context and form', *Casabella* 525: 16–25.

London Docklands Development Corporation (1982) *Isle of Dogs: a Guide to Design and Development Opportunities*, London: London Docklands Development Corporation.

Lotus International (1984) 'Steven Hall: Bridge of houses in Manhattan: typological variations on a rail structure, 1981', *Lotus International* 44: 41–3.

Lynch, K. (1960) *The Image of the City*, Cambridge, MA: MIT Press.

Maxwell, R. (1976) 'An eye for an I: the failure of the townscape tradition', *Architectural Design* 46, 9: 334–56.

Milton Keynes Development Corporation (1970) *The Plan for Milton Keynes*, Wavedon: Milton Keynes Development Corporation.

Ministry of Transport (1963) *Traffic in Towns*, London: HMSO.

Morris, E. K. and Levin, E. (guest editors) (1982) 'Typology in design education', special issue of *Journal of Architectural Education* 35, 2.

Pevsner, N. (1946) 'Visual planning and the City of London', *The Architectural Association Journal* 65 (December/January): 31–6.

Pevsner, N. (1947) 'The picturesque in architecture', *Journal of the Royal Institute of British Architects* 55, 12: 55–61.

Pevsner, N. (1955) *The Englishness of English Art*, The Reith Lectures. London: British Broadcasting Corporation.

Piper, J. (1944) 'Colour and texture', *Architectural Review* 95, 2: 51–2.

Piper, J. (1947) 'Pleasing decay', *Architectural Review* 102, 9: 85–94.

Portoghesi, P. (1958) 'Dal neo-realismo al neo-liberty', *Comunità* 65.

Price, U. (1794) *Essay on the Picturesque*, London.

Richards, J. M. (1941a) 'Towards a replanning policy', *Architectural Review* 90, 7: 38–40.

Richards, J. M. (1941b) 'Wanted: an hypothesis', *Architectural Review* 90, 11: 148–9.

Richards, J. M. (1942) 'A theoretical basis for planning', *Architectural Review* 91: 39–42, 63–70.

Richards, J. M., Pevsner, N., Lancaster, O. and Hastings, H. de C. (1947) 'The second half century', *Architectural Review* 101, 1, special supplement.

Rogers, E. N. (1959) Editorial, *Casabella* 228.

Rossi, A. (1966) *L'Architettura della Città*, Padua: Marsilio Editori. Translated 1982 as *The Architecture of the City*, Cambridge, MA: MIT Press.

Rossi, A. (1970) 'Due progretti', *Lotus International* 7: 43–8.

Rossi, A. (1975) 'Tipologia, manualistica architetture', in Bonicalzi, R. (ed.) *Scritti Scelti sull'Architettura e la Città 1956–1972*, Milan: CLUP.

Scolari, M. (1971) 'Un contributo per la fondazione di una scienza urbana', *Controspazio* 7/8, 3: 40–7.

Sharp, T. (1935) 'The English tradition in the town', *Architectural Review* 78, 11: 179–87.

Sharp, T. (1936) 'The English tradition in the town', *Architectural Review* 79: 17–24, 115–20, 163–8.

Silvetti, J. (1980) 'On realism in architecture', *Harvard Architectural Review* 1: 10–31.

Smithson, A. and Smithson, P. (1967) *Urban Structures*, London: Studio Vista.

Smithson. A. and Smithson, P. (1970) *Ordinariness and Light: Urban Theories 1952–1960*, London: Faber.

Vidler, A. (1968) 'Editorial', *Oppositions* 7: 1–4.

Vidler, A. (1978) 'The third typology', in *Rational Architecture Rationelle*, Brussels: Archive d'Architecture Moderne.

Watkin, D. (1982) *The English Vision: the Picturesque in Architecture, Landscape and Garden Design*, London: Murray.

Wild, D. (1977) 'British architecture: theory into practice', *Architectural Design* 47, 9: 591–4; 47, 10: 7–23.

Wilford, M. (1984) 'Off to the races, or going to the dogs?', *Architectural Design* 54, 1/2: 8–15.

Wittkower, R. (1952) *Architectural Principles in the Age of Humanism*, London: Tiranti.

Wolfe, I. de (1947) 'Townscape', *Architectural Review* 106, 12: 355–62.

7

THE EVOLUTION OF TWENTIETH-CENTURY RESIDENTIAL FORMS: AN AMERICAN CASE STUDY

Anne Vernez Moudon

INTRODUCTION

Massive suburban development over the course of this century has given the newer North American cities a form which differs significantly from that of earlier cities. Forty-five per cent of the US metropolitan population today lives in areas that are called suburbs. The country's metropolitan housing stock is new. Only 20 per cent of it predates the 1920s, although this figure varies from more than 40 per cent in the North East to less than 10 per cent on the West Coast. Piecemeal development characterizes the newer parts of cities, with only 10 per cent of residential suburban development being in the form of planned communities. Since the 1920s, residential development has occurred within a loose framework of local land-use laws, regulating the subdivision of land and the distribution of functional elements, and of building codes. Both land-use laws and building regulations support a strong tendency towards the ownership of single-family houses and automobile transportation. Over time, they have favoured tremendous increases in the spread of metropolitan areas. Today, the urbanized area of Los Angeles covers 474,000 hectares (1,170,780 acres) and has a population of 9.5 million people. In comparison, London's urbanized area spreads over 120,000 hectares (296,400 acres) and has a population of 6.5 million people. London's gross density of 56 persons per hectare (22.7 per acre) contrasts with a gross density of 20 persons per hectare (8.1 per acre) in Los Angeles. Though several of its parts are considerably older than those of Los Angeles, the New York–New Jersey urbanized area has a gross density of only 21.5 persons per hectare (8.7 per acre) (Newman and Kenworthy 1989). Such sprawling cities have morphological characteristics that differ substantially from European cities (Whitehand 1987).

Not surprisingly, US cities have generated their own set of problems. Traffic congestion infringes on productivity in the workplace. It threatens air quality as well as the quality of life in general – petrol consumption in US cities is four times higher than in European cities, and six times higher than in Asian cities. Lack of affordable housing, limited public services, and increased costs of infrastructure loom as further threats to urban living in the face of continued exurban fringe development (Downs 1989). In response, several states have recently adopted growth-management legislation, restricting development beyond urbanized zones and mandating that future growth take place within land that is already serviced. Such legislation marks a turning point in US urban design and planning practice: it forces local governments to stop relying on expansionary policies and to focus instead on existing urban systems.

This new, inward-looking climate in urban-planning practice raises the question of how existing urban systems can accommodate improvements in transportation, housing, public services and infrastructure. It demands better knowledge of these existing urban systems.

A large body of literature on suburbs has emerged in the last decade, reflecting the formidable expansion of North American cities during this century. Part of this literature deals with the historical forces that have shaped suburbs (M. P. Conzen 1980; Jackson 1985; Weiss 1987; Stilgoe 1988). Other studies monitor and assess socio-economic trends in suburban development, focusing on shifts in employment patterns, commercial activity, changing family structures, demographic and social conditions, and so on (Hayden 1984; Fishman 1987). But little has yet been said about the forms that have materialized: the houses, streets, roads, parks, schools, offices, parking lots and other elements that together weave the web of suburban forms. Although work on the history of the house is developing rapidly, providing much-needed information on typical forms of shelter (Gowans 1964; Handlin 1979; Foley 1980; G. Wright 1981; McAlester and McAlester 1984), it generally does not include more recent house forms, nor does it relate house forms to their host districts. Further, ample documentation exists on the form of model suburbs (H. Wright 1935; Stein 1971; Stern 1981; Hayden 1984). Yet if these carefully designed, often innovative suburban forms have served as exemplary precursors to later development, they, in effect, bear little resemblance to the mass of developer-controlled subdivisions whose standard designs remain selective interpretations of their famous antecedents.

This chapter attempts to identify significant stages in the evolution of common (as opposed to exemplary) residential suburban forms in North America since the 1920s. It examines the forms of existing suburban residential areas, the prevailing design practices, and their evolution during this century. In the last seventy years, suburban building has undergone significant changes. The ubiquitous single-family, detached, suburban house for those in the middle-income brackets has given way to semi-detached houses, terrace houses of three or four dwellings, and apartment blocks. More than 40 per cent of the new housing built in North American suburban areas is now in the form of multiple units. Furthermore, curved streets and culs-de-sac have replaced the traditional gridiron street pattern. In this chapter, consideration is given first to the important elements of suburban residential form, namely the properties (houses on their lots) and the street networks. An outline is provided of a typology of basic, common housing forms which reflect specific stages in the evolution of suburban development and their corresponding land development and building traditions. The focus is on middle-class housing. The second part of the chapter illustrates how these basic forms have been used in specific cases of residential environments in Seattle's Puget Sound. The third part of the chapter identifies the trends that have emerged in the design of street networks, and house and lot forms, and in the relationships between open and built space, and the characteristics of private and public open space. A discussion highlights the impact of these trends on the provision of housing and transportation. The chapter concludes with suggestions for the future management of suburban landscapes.

A TYPOLOGY OF COMMON RESIDENTIAL FORMS

House forms, lot sizes and street layouts define the essential elements of urban and suburban form (for widespread agreement on the primacy of these elements in morphological analysis, see M. R. G. Conzen 1968; Panerai *et al.* 1980; Caniggia 1983; Moudon 1986a, forthcoming). Certain house forms usually correspond to particular lot sizes. This is especially true in middle-income housing where the cost of land is an important component of the total cost of development and where, as a result, the lot is only large enough to accommodate the house comfortably. Lot forms are in turn influenced by street patterns: for instance, curved streets entail non-orthogonal lots. In given areas, streets, lots, and houses combine to form 'plan units', a term identified by M. R. G.

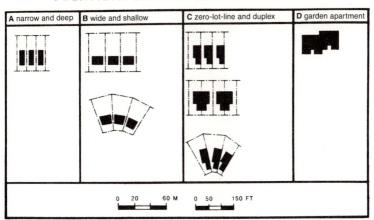

Figure 7.1 Basic types of houses and lots.

Conzen to distinguish between areas of different morphogenetic plan types. Plan units are socio-architectural units of developed land which encapsulate the social ideals and economic limitations of the first builders and residents (M. R. G. Conzen 1968).

Four basic types of house forms are identified which depict the different generations of North American suburban habitats between the 1920s and the 1980s (figure 7.1). These house forms relate to specific lot sizes and shapes, which reflect the land subdivision pattern. Architectural style is not incorporated in the typology.

Three basic types of street networks have structured movement patterns and the related land subdivision during the same period of time (figure 7.2). Combinations of basic types of houses and lots and street networks create basic types of plan units which characterize entire suburban areas. Three basic plan units have been identified (figures 7.3, 7.4 and 7.5).

The typologies of houses, lots and street networks are further described below. These descriptions draw on several years of research covering a variety of residential districts in different North American cities and offer a theoretical framework for discussion of the potential of these forms and to guide future, more specific, research on existing residential developments. The synthetic nature of these typologies stems from the fact that suburban residential forms present special challenges for morphological analysis: the enormous size of recent metropolitan development demands substantial mapping capabilities, and the poor quality of existing information on the characteristics of urban form

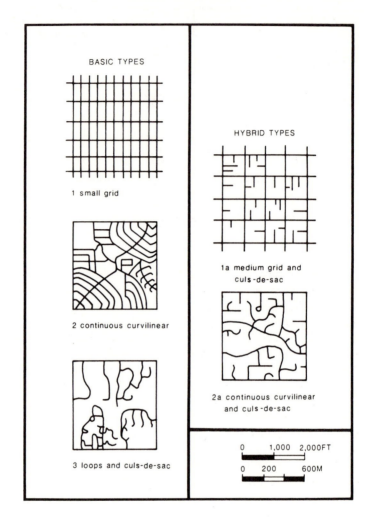

Figure 7.2 Basic types of street networks.

necessitates extensive field studies. In the absence of substantial budgets to support this kind of research, systematic morphological analysis can only be carried out on discrete, limited parts of metropolitan areas. Such analyses will necessarily be limited in space and time and will lack the broad-based approach required for the development of long-range city-planning policies. In this light, the typologies proposed here represent a first step towards organizing the study of suburban residential forms.

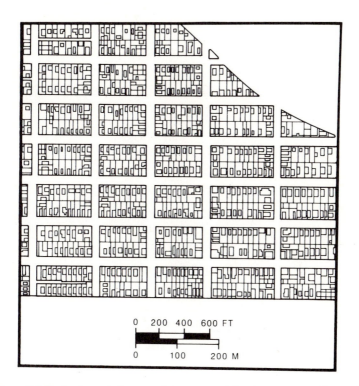

Figure 7.3 Basic plan unit. Lot type A (narrow and deep), street-network type 1 (small grid).

Although they need verification, they should be paradigmatic for communities looking to improve themselves.

Basic house and lot types

The *narrow-and-deep* single detached house (type A, figure 7.1) occupies a lot of 40 or 50 feet by 100 feet or more (12.2 or 15.2 by 30.5 metres).[1] Common until the 1930s, this type is used only selectively today.

The *wide-and-shallow* single detached house (type B, figure 7.1) occupies a lot of 60 feet by 100 or more feet (18.3 by 30.5 metres). Originating in the 1940s in the form of *ranch* and later *split-level* houses, the type remains prevalent today, but in medium-priced, single-family neighbourhoods. The geometry of the lot varies according to the street network.

175

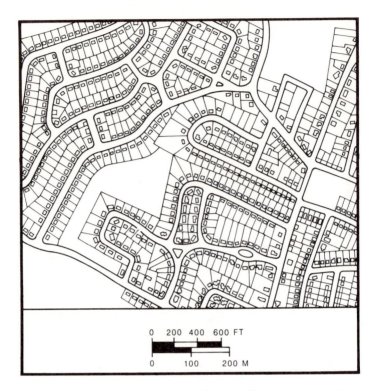

Figure 7.4 Basic plan unit. Lot type B (wide and shallow), street-network type 2 (continuous curvilinear).

The *zero-lot-line* house (type C, figure 7.1) occupies a lot of 40 or 50 feet by 100 feet (12.2 or 15.2 by 30.5 metres). With origins in the late 1960s, this type can be combined to create semi-detached houses (double houses or new forms of duplexes). It is found primarily in planned developments featuring common open space and other shared amenities.

The *garden apartment* (type D, figure 7.1) is part of a development which can house from as few as ten dwelling units to more than a hundred such units. Hence the land associated with this type can vary greatly in size. Although the building type is not new, its generalized use in large-scale suburban projects dates from the 1970s.

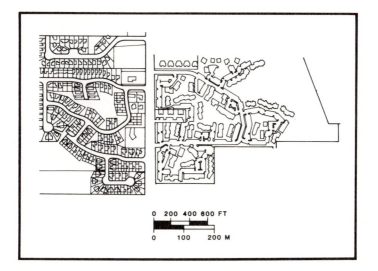

Figure 7.5 Basic plan unit. Lot types C and D (zero-lot-line, garden apartments), street-network type 3 (loops around culs-de-sac).

Types of street networks

The *small gridiron* of streets (type 1, figure 7.2), measuring 260 to 360 feet by 460 to 660 feet (79.2 to 109.7 metres by 140.2 to 201.2 metres),[2] is a type which prevailed until the 1930s, but is no longer in favour today.

A network of *continuous curvilinear* streets is a type (type 2, figure 7.2) which emerged in the 1930s as a reaction to the perceived monotony of the small gridiron. With origins in the mid- to late-nineteenth-century romantic suburb (used, for example in F. L. Olmsted's Chicago suburb of Riverside), this type only became standard in the late 1930s, as the newly created and influential Federal Housing Administration promoted its virtues. West of the Ohio Territory, subdivisions of curvilinear streets are often bounded by the traces of the 1785 Land Survey. Thus large, 2,640-foot (804.7-metre) square grids of former agricultural roads, corresponding to the quarter section of the Land Survey,[3] serve to provide a network of straight, continuous arterial streets for circulation.

While the continuous curvilinear street network offers the potential to adapt to topographical changes and, as such, can be an improvement over the small gridiron, its standard application often ignores the

177

particulars of the terrain. The type ceased to be popular in the 1960s.

A *loop road* feeding the subdivision (type 3, figure 7.2) has prevailed since the 1970s. It is an adaptation of the Radburn garden city model, which sought to mitigate the nuisance and dangers associated with increasing automobile traffic. The loop road can assume a variety of shapes. Usually, however, it has no more than two points of access to arterial through-streets, and hence excludes traffic which is not directly related to the internal needs of the subdivision. Few, if any, houses in loop subdivisions face directly on to the loop. Instead, they are served by culs-de-sac which complement the subdivision street system. As in the previous type, this street network is often inserted in the large agricultural grid of the Land Survey.

Several factors create *hybrid street types* (types 1a and 2a, figure 7.2). First is the presence of medium-sized grids intended originally for small agricultural production at the edge of late-nineteenth- and early-twentieth-century cities. These 660- to 1,320-foot grids (201.2 to 402.3 metres) were subdivided into lots too small to satisfy the needs of the developers of large-scale 1950s subdivisions who preferred large, and comparatively inexpensive, areas of land beyond the then urban fringe. Thus land in the medium-sized grids was left to mature gradually over time and, as such, presented special conditions. While some of the contemporary suburban agglomerations exhibit similar street patterns today, these conditions do not appear to be sufficiently common to warrant classification as a basic type of street network.

The cul-de-sac, used as a minor street to promote safe and quiet residential environments, is the second factor modifying the basic street types. An age-old device to reach the interior of city blocks with new development, culs-de-sac have also been common in urban fringe areas where deep lots have eventually been subdivided into 'flag lots' – back-to-back lots in the shape of a panhandle. However, while until the 1930s culs-de-sac were used primarily as a technique to remedy anachronistic land patterns, since then they have gained in popularity for the platting of new subdivisions. They are typically found in medium-sized gridirons, in late curvilinear street networks, and in loop arrangements (Relph 1987: 52, 174).

Plan units

Combinations of basic types of houses, lots and street networks create plan units or areas of different morphogenetic composition and character. Plan units generally represent areas developed or built within a

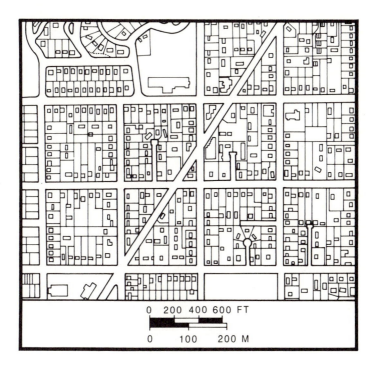

Figure 7.6 Hybrid plan unit. Lot type B (wide and shallow), street-network type 1a (medium grid and culs-de-sac).

limited period of time by one entity, or several entities with similar intentions. For the purpose of this classification, a distinction is made between *basic plan units* which include house, lot and street types of a given period, and *hybrid plan units* which combine house, lot and street types of different periods.

Basic plan units synthesize development practices that are characteristic of a given period. Together, they identify different generations of such practices. First, small gridirons used in junction with narrow-and-deep houses and lots (basic types 1 and A) are most common until the late 1930s. Secondly, curvilinear through-streets used in conjunction with wide-and-shallow houses and lots dominate in the 1940s and the 1950s, but continue into the 1960s (basic types 2 and B). Thirdly, loop roads are most common in current development, used in conjunction with several house and lot types, including the wide-and-shallow house and lot, the zero-lot-line house, and the garden apartment (basic types 3

179

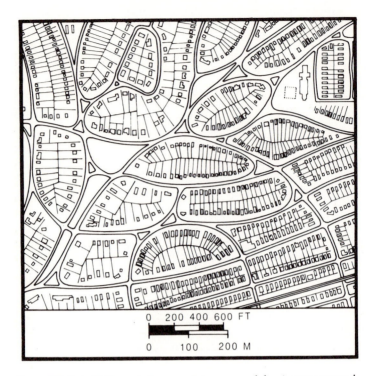

Figure 7.7 Hybrid plan unit. Lot type A (narrow and deep), street-network type 2 (continuous curvilinear).

and B, C, and D). Thus the use of loop roads corresponds to a new level of maturity of the suburb: whereas early suburbs have one basic type of single-purpose house form, contemporary ones introduce several types of houses and lots, responding to different kinds of people and incomes, and they are thus beginning to project the formal complexity that is usually found in cities.

Examples of hybrid plan units are shown in figures 7.6, 7.7 and 7.8. Such plan units identify development practices which mix house, lot and street types of different periods. The reasons for mixing these types can be many. Generally, areas where development took place over several decades are likely to be hybrid plan units because they will host several generations of houses. Areas structured by a hybrid street system will by definition constitute hybrid plan units. Further, the use of innovative forms of street networks or house designs, or, conversely, the revival of old forms, will also create hybrid plan units.

180

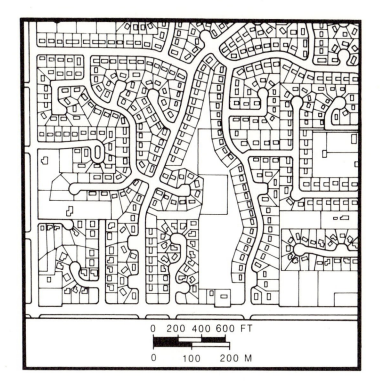

Figure 7.8 Hybrid plan unit. Lot type B (wide and shallow), street-network type 2a (continuous curvilinear and culs-de-sac).

APPLICATIONS OF THE BASIC TYPES: CASE STUDIES IN THE PUGET SOUND

What is the reality of the basic morphological types discussed so far? How helpful are they in analysing the formal characteristics of suburban environments? Case studies selected from a range of suburban conditions in Seattle's Puget Sound area illustrate specific applications of the types presented.

Wallingford: from the 1920s to the 1990s

Wallingford is a case in which several nineteenth-century suburban plats matured into a residential community in the late 1920s. It is a clear example of the first basic plan unit described above (types 1 and A).

181

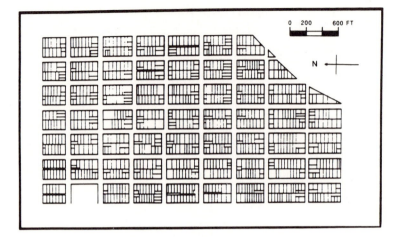

Figure 7.9 Wallingford: streets and lots.

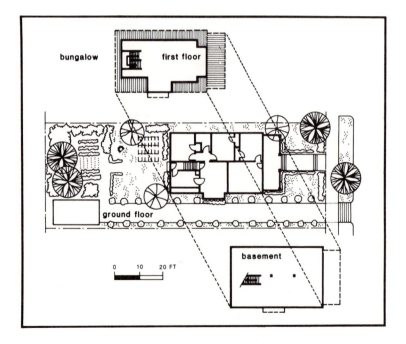

Figure 7.10 Wallingford: typical house and lot.

Figure 7.11 Wallingford bungalows built in the 1920s (photograph by the author).

The site was a logging area where, in the late 1870s, four families set up farms on the hill sloping down to Lake Union. The area was never an independent city, as were some of the neighbouring settlements, but served as a residential extension of these small cities. It was annexed by Seattle in 1891. Most of Wallingford was platted by 1889 (Historic Seattle Preservation and Development Authority 1975). Located on rail lines, the area saw first a mill move in, and then, at the turn of the century, a gas works. Well serviced by public transportation, the residential part of the neighbourhood first grew between 1910 and 1920. A playing field was in existence in 1925, when only half of the area was developed with houses. By the time of the Depression, however, Wallingford was almost entirely built up with single-family houses (Housing Authority of the City of Seattle 1940; Heasly 1986).

Wallingford was platted with blocks running north–south and measuring 240 feet by 360, or 400, even 480 feet (73.2 metres by 109.7, 121.9, or 146.3 metres). Street widths range between 53 and 66 feet (16.2 and 20.1 metres) in the east–west direction (North 36th Street being 53 feet (16.2 metres), North 35th Street being 60 feet (18.3 metres), and the others being 66 feet (20.1 metres)). They range between 60 feet (18.3 metres) and 66 feet (20.1 metres) in the north–south direction, North Meridian Avenue being an example of the smaller width and Burke Avenue being an example of the larger width. Streets and blocks represent the first basic type (type 1) (figure 7.9).

Most of the area was initially platted with lots of 60 by 120 feet (18.3 by 36.6 metres), although some lots were as narrow as 25 feet (7.6 metres). However, the majority of lots are now 40 feet (12.2 metres) wide, because land was in such demand in the 1920s that many owners combined two 60-foot (18.3 metres) lots into three 40-foot lots. The various platting and land-buying decisions created lot sizes that today vary from 4,000 to 7,000 square feet (372 to 651 square metres).

Most houses were built individually between 1910 and 1930. They represent the basic narrow-and-deep type (type A). While different architectural styles emerged over the course of the development of the neighbourhood, the bungalow style prevails in some 60 per cent of the structures (figures 7.10 and 7.11). Houses typically measure 25 by 40 feet (7.6 by 12.2 metres), have a full basement, and a partially habitable attic. Originally, families lived in some 2,000 square feet (186 square metres) of finished space. Today, this space has often been expanded to 2,800 square feet (260.4 square metres). A small, single-car garage is tucked into the back of the lot or, in houses built from the 1920s onward, the garage is incorporated into the basement of the house.

Churches and small grocery stores initially dotted the different parts of the neighbourhood, with an important commercial centre eventually emerging along North 45th Street. Over time, some of the single-family structures have been converted to two-family units or duplexes. Others were demolished and replaced by modern houses or even apartment houses. Since the late 1970s, however, the area, with the exception of the commercial strip and some commercial pockets, has been designated as a single-family-only zone. Multi-family accommodation has therefore been 'grandfathered' in the zoning code – meaning that non-single-family uses existing before rezoning have been legalized, but subject to restrictions if the structures are remodelled or modernized.

Haller Lake: from the 1950s to the 1990s

Haller Lake is an early-twentieth-century ex-urban farming area that has evolved into a residential community. It has been gradually developed, primarily with single houses, since the 1950s. Annexed in the early 1950s to the City of Seattle, it is experiencing full maturity only today. Haller Lake illustrates a hybrid plan unit (types 1a, A, and B) (figure 7.12).

The area was originally platted with large, 600-foot (182.9-metre)

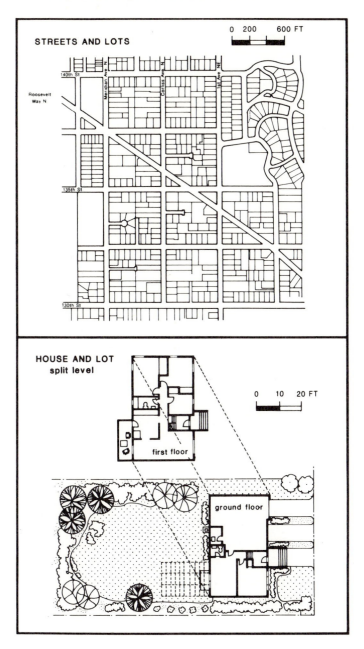

Figure 7.12 Haller Lake: morphological characteristics.

185

square blocks and 60-foot (18.3-metre) wide streets, a module fitting into the mile-square grid of the Land Survey, and often used at the edges of towns to accommodate small farms. This layout corresponds to the hybrid, medium-sized grid type (type 1a). Haller Lake was developed over a longer period of time than Wallingford, and thus exhibits a greater variety of house types (figure 7.13).

The original lots measured 120 by 300 feet (36.6 by 91.4 metres). A few farmhouses were built at first, usually occupying the corners of the blocks. By the 1950s, the land was further subdivided into 60-foot (18.3-metre) residential lots of 6,000 to 10,000 square feet (558 to 930 square metres). Small houses were then built along the streets, leaving large expanses of open land in the middle of the large blocks (a common pattern of gradual development of agricultural land – see Tunnard and Pushkarev 1981: 81). The infilling of the central parts of blocks began

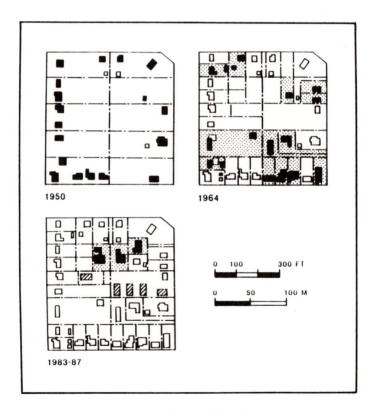

Figure 7.13 Thirty years of developing Haller Lake: a sample street block.

186

Figure 7.14 Haller Lake split-level house built in the 1960s (photograph by the author).

to be common in the 1960s, and continues today. It takes the form of flag lots, to gain private access to the street. As many as four flag lots with private alleys can be found within the 300-foot (91.4-metre) depth of the original lot. Few flag lots actually share road rights-of-way. Alternatively, small developers have laid out culs-de-sac serving lots on sometimes one, but generally both, of their sides.

Most main streets are 60 feet (18.3 metres) wide. The added small streets, alleys, and culs-de-sac fit orthogonally within the original block; they may or may not have been dedicated as public rights-of-way.

Some of the original small farmhouses and barns are still evident today here and there in the neighbourhood. There are also a few pre-1940s bungalows of the same type as found in Wallingford (type A). But most houses are of the era immediately following the Second World War. They are typical ranch or split-level houses on lots 60 feet (18.3 metres) wide (type B) (figure 7.14). Development in the back of the blocks used the same building types until the 1970s, when larger houses began to be built, reflecting the fact that by then only upper-middle-class segments of the population could afford such housing types. The basic ranch houses of the 1950s typically measure 45 by 20 feet (13.7 by 6.1 metres), affording 900 square feet (83.7 square metres) of habitable space. They are often flanked by a car-port. Split-level houses have the

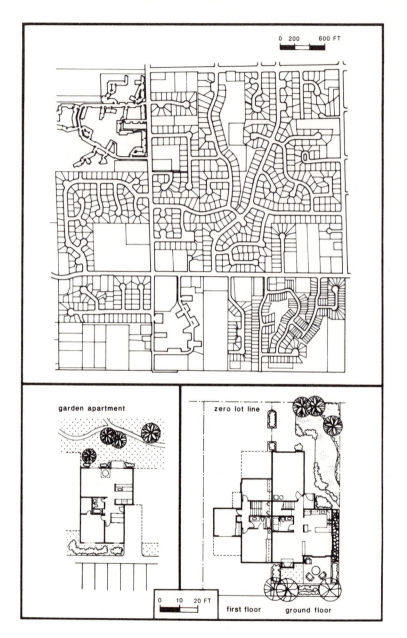

Figure 7.15 Unincorporated King County and Kirkland: morphological characteristics.

188

same floor plate as ranch houses, but are two storeys high, typically affording 1,800 square feet (167.4 square metres) of habitable space, including a one- or two-car garage. Recent versions of split-level houses are somewhat larger: they may remain 45 feet (13.7 metres) wide if they are to fit into 60-foot (18.3-metre) lots, but they are deeper than their antecedents, ranging between 25 and 35 feet (7.6 and 10.7 metres), and augmented by different types of wings or small protrusions.

Unincorporated King County and Kirkland: from the 1970s to the 1990s

Unincorporated King County and Kirkland is a case where land was surveyed and platted for development in the latter part of the nineteenth century. However, agricultural uses prevailed for almost a century, maintaining the coarse land subdivision contained within the government survey. Since the 1960s, the agricultural division of land has served as a framework for a patchwork of different types of residential development. The mile-square grid continues to govern the spatial structure, and through-streets occur fairly regularly on the quarter section, every 2,640 feet (804.7 metres). As a general pattern, the 160-acre (64.8-hectare) quarter section has been divided into four 40-acre (16.2-hectare) parcels which, in turn, have also been subdivided into four 10-acre (4.1-hectare) parcels.[4] However, in many instances, parcels smaller than 10 acres can be found, creating a somewhat irregular pattern.

Small farms and barns, though still visible in the landscape today, are quickly being replaced by residential subdivisions which can be classified into three different types. Subdivisions of single-family, detached houses on curved streets with culs-de-sac began in a piecemeal fashion in the early 1970s and are continuing today. More prevalent today, however, are small planned unit developments of semi-detached houses (some are triplexes and fourplexes), and garden-apartment complexes (figure 7.15).

Single-family subdivision in the 1970s

The Firlock subdivision illustrates a hybrid plan unit (types 2a and B). Curvilinear streets and culs-de-sac are inserted within the quarter section, leaving a few parcels which were already built upon. The few through-streets and the numerous culs-de-sac make this area an example of a hybrid street type (type 2a). They provide access to irregularly shaped lots which are 50 to 80 feet by approximately 100 feet (15.2 to 24.4 by 30.5 metres) – ranging from 6,000 to 8,000 square feet (558 to 744 square metres). This particular subdivision falls just short of 500 lots.

Houses are built according to the developer's standard, and houses and lots are sold together to individual owners. Split-level houses predominate, providing a range of architectural styles that gives variety to the area. These houses belong to the same basic morphological types as the 1950s houses found in Haller Lake (type B), but the floor plans are larger, as are the internal spaces (figure 7.16). Increased size is accommodated by protrusions from the basic box (creating, for instance, L-shaped buildings). These protrusions are deemed to enhance the house stylistically. Including a two-car garage, most houses provide some 3,000 square feet of living space (279 square metres).

Planned unit development in the 1980s

The Upland Green community is a planned unit development (PUD) which illustrates the third basic plan unit (types 3 and C). Part of a three-project experiment by King County to provide affordable housing, the project received a building permit in 1980.

Upland Green accommodates 150 dwellings on 11.8 hectares (29.25 acres), with 4 hectares (9.95 acres) reserved as common open space. The developer was able to tap county services from the main road, but had to build a sewer connection almost half-a-kilometre long within the site. All streets within the project have been dedicated to the county.

Organized as a loop around the site feeding a series of culs-de-sac (type 3), the street network is based on a 40-foot (12.2-metre) right-of-way, with a 24- or 28-foot (7.3- or 8.3-metre) paved area. Only two streets have a sidewalk on one side. All other pedestrian ways are in the form of trails away from the streets.

Lot sizes are 4,000 and 5,000 square feet (372 and 465 square metres). All houses are a variation of the zero-lot-line type: a house with three exposures and a blank wall within a few feet of one of the side lot lines. This basic morphological type (type C) can take the form of a single-family detached house, or the two blank walls can be joined to make a double house for two or four families (figure 7.17). Including the two-car garage, the single-family house provides more than 2,000 square feet (186 square metres) of space. Semi-detached houses (duplexes) can provide the same amount of space, or less depending on the variation used. Terraces of four dwellings (fourplexes) offer some

Figure 7.16 Firlock's split-level homes built in the 1970s (photograph by the author).

1,000 square feet of space (93 square metres). The relatively generous front yards make the back yards small, barely accommodating a patio and a lawn.

Garden apartments in the 1980s

These developments take place in variations of the 10-acre (4.1-hectare) parcel or a quarter of the quarter section. They are yet another version of the third basic plan unit (types 3 and D). Salish Village (168 units), Falcon Ridge (102 units), Arrowood (94 units) and Shawnee Village (160 units) are adjacent garden-apartment condominiums built over the past twelve years. The four developments fit into a quarter section which also includes a hospital site in its south-western portion. They are fed by a shared internal loop road (type 3), which is lined with parking stalls and carports. Buildings accommodating one- to three-bedroom apartments are scattered along the loop road, separated from the parking area by a narrow band of green spaces. At the back, common green open space separates the buildings (figure 7.18). A large common green space borders the site to the south; on a steep slope, this space is not directly usable, nor is it feasible for development at this time. Buildings are three storeys high, and all apartments are reached by

Figure 7.17 Upland Green: zero-lot-line houses built in the 1980s. A pedestrian trail to the common green is in the foreground (photograph by the author).

stairs serving two apartments per floor (type D). Condominium apartments offer 900 to 1,200 square feet (83.7 to 111.6 square metres) of habitable space, excluding the garage. Ground-floor apartments have a small patio on the back green, and upper apartments have a small balcony above the patio.

TRENDS IN THE EVOLUTION OF BASIC MORPHOLOGICAL TYPES

These case studies show that suburban residential form is anything but a monotonous continuity of shapeless elements. Compared to earlier urban residential forms, these suburban forms can appear loose and somewhat disconnected, with many, often small, singular structures spread across the landscape and much undefined space left vacant – open space being an undeserved euphemism in suburban environments. Yet different types of forms are clearly identifiable. Furthermore, a marked evolution can be detected which signals an increase in formal complexity. It suggests that the early, simple suburban forms were a mere first step in an evolutionary process that resembles its urban coun-

Figure 7.18 Shawnee Village garden apartments built in the 1980s (photograph by the author).

terpart. Several specific trends in the evolution of twentieth-century residential forms require discussion.

The increasing width of houses and lots

Throughout the 1950s, the single detached house dominated the suburban, middle-class landscape. Yet the form and organization of the single house changed so radically during the first half of the twentieth century that two types can be identified. Until the 1930s, a typical 40-foot lot was occupied by a narrow-and-deep house. Following an established residential layout, service functions such as kitchens, bathrooms and laundry facilities were relegated to the back of the house and lot. The house had at least a partial basement for service activities, which also occasionally housed an autonomous family member or a tenant. The attic sometimes contained finished rooms. When the private car first emerged as the prize possession of many families, it was parked in the back of the lot in a newly built garage or car-port. The rest of the backyard was grassed, with perhaps a vegetable garden or a flower garden. The front yard was the public face of the property, where the house often exhibited a porch for late afternoon and Sunday leisure. Front-yard design remained simple and singular, with a lawn and a few

193

shrubs along the pathway and at the foot of the house, to hide the foundations. If, because of topographical constraints, the car could not be brought to the back of the lot, a garage was built on the street at street level, often buried under the elevated front yard. In the 1920s, the garage became incorporated in the house, notably in the basement, and was accessible directly from the street. This was an important stage of mutation in the type since it brought a service function, individual transportation, to the erstwhile-ceremonial side of the house – whose plan, however, would remain unchanged for a few decades.

Second, in the 1940s and especially in the 1950s, a typical lot was at least 60 feet (18.3 metres) wide and was occupied by a house which was not larger, and indeed was sometimes smaller, than before. Because this second basic type of lot was considerably wider than its predecessor, the house form and organization were turned through 90 degrees. Thus houses were wide and shallow. The ranch or split-level house was one or one and a half storeys tall, reflecting the smaller size of families. Both kitchen and garage lined the street, which became a functional and service space. Garage space took the form of a car-port at the side of the house, or was incorporated in the basement at the front of the house. It was wider than in the 1920s, because cars were larger, and sometimes it could accommodate two cars. The front yard now contained a small decorative garden and an often substantial asphalted area for the driveway. Cleared of functional elements, the back yard became a private place for the family to use.

The switch from a narrow-and-deep to a wide-and-shallow single-house and lot type has not been studied extensively. At least two circumstances would seem to explain the phenomenon. First was the emergence of the private automobile. Because the car in effect reduced the distance between activities, the lot could become wider, and community activities more spread out in the landscape. Wide lots and houses were signs of wealth which developers wanted to exploit. Furthermore, housing advocates and planners were critical of the dark interiors of narrow and deep houses, denigrating the side yards as useless spaces which should be eliminated. Wide-and-shallow houses resolved these problems by reducing the length of side yards and providing needed interior light (H. Wright 1935; Moudon 1986b).

The return to narrow houses and lots

The wide-and-shallow house type is still used today, but it is now aimed at the upper-middle-class market. The middle class, if able to afford a home, is likely to move into a planned development of zero-lot-line houses. In these developments, which have prevailed since the 1970s, the lot is smaller, to make the house affordable and to reduce maintenance costs. But the front yard has remained relatively deep, the back yard having been reduced to a minimum. Residents have access to a new type of open space: the private, but shared common green. The houses are again narrow, to use the land and the infrastructure more efficiently. The garage remains incorporated in the front of the house, and, as a double-car garage, it now occupies most of the front of the house. Increasingly, it is used for storage, while cars stay on the streets. The increasing presence of cars, parked on streets or stored in endless rows of garages, transforms the function of the street by eliminating its social dimension. With only cars on the street, accidents increase and pedestrians decrease in numbers.

In the 1980s, the middle-income people increasingly turned to the affordable contemporary garden apartment. Maintenance of this type of housing is low, but residents also see their private open space reduced to a minimum. The common green area has become a visual amenity, which cannot be used for active recreation.

Increased control of the development process

Until the 1930s, most developers were primarily land subdividers (Weiss 1987). Serviced but unimproved residential lots in gridded districts were sold directly to future residents, who then had to hire contractors to build the houses. Since the 1950s, developers have tended to expand their products, subdividing and servicing the land, and improving it by commissioning builders to construct different models of houses which are then sold to residents. Thus large areas are developed at the same time with similar lots and houses.

Large, common open space traditionally remained separate from the subdivision; it was the responsibility of counties or cities which reserved or bought land whenever required by the scale of private development. From the 1960s onward, however, developers furthered their control over the planning and design of the community, packing houses together tightly to make room for common open space within the development. Unlike its predecessors, however, this open space was privately held. The planned unit development (PUD) was becoming normal.

Planned unit developments

Since as early as the 1960s, new planning controls have encouraged developers to provide large open spaces and recreational facilities for the residents in exchange for higher densities in the residential development. A distant cousin of the garden-city concept, the PUD, was born. It had common green space in the centre of the residential district, and its pedestrian trails were separated from the streets, which were now the exclusive domain of the automobile. Streets usually looped around the development, serving houses on both sides. Lots were narrow again (40 feet (12.2 metres)), but also shallower (100 feet (30.5 metres) or less), and the fronts of houses were dominated by two-car garages. The front yard remained, but the back yard became small and private. Individual gardening was kept to a minimum, and common spaces were maintained professionally.

In PUDs, developers form a homeowners' association, which they control until all properties are sold. By-laws regulate the physical appearance of houses, gardens, streets and other common open spaces to ensure the marketability of the subdivision. The homeowners' association uses the by-laws to manage the private streets, common spaces and other changes requested by residents as time passes.

The intellectual origins of PUDs can be traced back to the English square, the eighteenth- and nineteenth-century English housing estates, and the New England Common Green. Specific US precedents include Louisburg Square in Boston's Beacon Hill, which was managed by a homeowners' association in 1844, and the New England mill towns (themselves inspired by European industrial new towns such as Port Sunlight). Yet much of the modern groundwork for PUDs in the US emanates from the 1920s efforts of American garden-city proponents Henry Wright and Clarence Stein, with the support of Lewis Mumford and other prominent planners and critics of suburbs just after the First World War. Actual revisions in zoning ordinances to encourage PUDs appeared after the Second World War, as, for instance, in 1949 in Maryland, 1950 in Pittsburg, 1952 in Alexandria, and 1962 in San Francisco. An early definition of PUDs was given by Babcock *et al.* in 1965 (Burchell 1972: 34), and the model ordinance used by Burchell is East Windsor's in New Jersey, adopted in 1967 (Burchell 1972: 15). There were 22,000 PUDs in the US in 1976 (Perin 1977: 105). Exactly why these forms of development become prevalent only in the late 1960s and early 1970s is not clear.

The principles behind PUDs are single ownership of a fairly large development (many jurisdictions set it at a minimum of 10 acres (4.1 hectares)) and co-ordinated collective management combined with private ownership. PUDs undergo so-called master planning, relating to a single-permit process; mixed uses, meaning at least mixed residential types (there are differences between the planned residential development and the PUD, the latter usually including uses other than residential); and arrangements for density transfers. Perceived advantages are unity of style; single permits (hence time savings); economies of scale, especially in the infrastructure; access to amenities; and the provision, at reasonable cost, of both basic facilities, such as schools, parks and recreational centres, and special facilities.

Residential densities

Residential densities differ substantially between the different types of houses, lots and street layouts. Densities considered here are those found in the case studies. Thus Wallingford represents the small grid and the narrow-and-deep house, or the 1920s basic plan unit; Haller Lake represents the medium-sized grid and the wide-and-shallow house, or a 1950s hybrid plan unit; Upland Green represents the loop road and the zero-lot-line house, or the 1970s PUD plan unit; and Kirkland's apartments represent the loop and the garden apartment, or the 1980s garden-apartment plan unit.

As defined in this study, *net densities* provide an indication of the influence of house and lot type on density, while *gross densities* describe the impact of street networks and private, but shared, open space (public parks and unusable private shared open space have been excluded from the measure of gross density) (table 7.1).

Net densities decreased by approximately 50 per cent between the 1920s and the 1950s, corresponding to the increased size of individual lots. Yet densities increased by more than 100 per cent between the 1920s single-family houses and the 1980s garden apartments. The range of densities yielded by single-family development is broad, a fact that is often not understood by lay communities, who attach a lot of importance to the land-use types but do not address specific densities within the type.

PUDs have a relatively high net residential density. Yet their gross density is relatively low, because as much as 40 per cent of their area consists of common open space. PUD residents often complain about houses being too close together, and feel ambivalent about the relative value of the common open space. The amount of space for streets and

Table 7.1 Changes in physical forms and residential densities.

	1920s	1950s	1980s	1980s
House and lot types	Single family	Single family	Planned unit development	Apartments
Street types	Small grid	Medium grid	Loop	Loop
Net density[1]				
Units per acre	8.75	4.36	10.7	17.3
Units per hectare	21.88	10.9	26.75	43.25
Gross density[2]				
Units per acre	6.78	3.72	5.03	13.2
Units per hectare	16.95	9.3	12.58	33.0

Notes: 1 Excluding streets, common green and private recreational facilities
2 Including local streets and private recreational facilities

private open space is similar in the 1920s single houses and the 1980s garden-apartment types – these spaces occupy between 15 and 24 per cent of the total area. It should be noted, however, that the amount of space allocated to internal streets in garden apartments is deceptively large since this space also includes private parking. In the other plan types, parking is included in the net density figure because it belongs to the private lot. Not surprisingly, the medium-sized grid has the smallest amount of street space of all plan types.

Housing densities must also be considered at the city and the regional scales. What happens, for instance, to these densities in large areas of contemporary development where 1950s single-family types, PUDs and garden apartments are frequently mixed?

The ratio between single-family dwellings and apartments does not exactly correlate with housing densities. The older cities of Seattle, Tacoma, Everett and Bremerton have developed with a large percentage of their housing as single-family homes. The newer, suburban cities have seen a lot of recent apartment development. The ratio of single-family dwellings to multi-family dwellings is the same in many suburban cities as that found in the older cities. It is possible to have net densities below five units per acre in suburban areas containing a large percentage of apartments if the single-family dwellings average no more than two units per acre (in fact, newer suburban areas in the Puget Sound have net residential densities below five units per acre). Similarly, in older cities, it is possible to have high net residential densities with a large percentage of single-family housing if the single-family dwellings are built on small lots – older suburbs in the Puget Sound yield five to nine units per acre.

(Puget Sound Council of Governments 1990: 86)

Densities have important repercussions on the availability of, and access to, city services. It is generally accepted that reasonably regular (approximately every twenty to thirty minutes) public transport (by bus) can be provided in densities of at least 17.3 dwelling units per hectare (7 dwelling units per acre, a gross density figure as defined in this study). Indeed, most newer suburban areas are not well served by public transport. Similarly, gross densities of 22.2 units per hectare (9 units per acre) extending over areas of 10,000 or more people can provide an acceptable pedestrian environment, allowing residents to walk to convenience shops, schools and recreational facilities.

From grids, to curves, to loops: the death of the public street

Street networks have changed radically between the 1920s and 1980s. In the 1930s, the ubiquitous grid was replaced by curved streets as a popular device to plan residential districts. Advocates of curved streets praised their picturesque qualities as well as their ability to meet all kinds of topographical conditions (Federal Housing Administration 1938). In reality, however, few curved-street subdivisions seem particularly ingeniously integrated into their terrain. Moreover, as residential streets were increasingly, from the 1960s, laid out to discourage through traffic, they looped around the development with often only a single access point to a feeder arterial.

The trends are clear. First, the undifferentiated grid of the 1920s gave way to a hierarchy of streets: arterials, which often remained as part of a grid, were complemented by residential streets which became private lanes. While early suburbs were gridded with 60-foot (20-metre) rights-of-way every 200, 400 or 600 feet (61, 121.9 or 182.9 metres), depending on the direction of the grid, it was not unusual for residential areas developed after the 1970s to have public rights-of-way at intervals of only 2,640 feet (804.7 metres). This explains the severe traffic congestion experienced in new suburban areas, as all residents must eventually get to the few arterials which provide through access to other areas of 'town'.

Using the plan units described in the density studies, there were the following lengths of public through-street per dwelling: in the 1920s, 12.4 metres (40.7 feet); in the 1950s, 6.6 metres (21.8 feet); in the PUD, 4.9 metres (16.1 feet); and in the garden apartments, only 1.9 metres (6.2 feet). Translated back into the gridded suburb of the 1920s, 1.9 metres of public through-street would mean a gross density of 109.9 dwellings per hectare (44.5 dwellings per acre), or more than six times the existing densities in the Wallingford case study. Given that suburban areas now have almost as many cars as they have inhabitants, today's traffic congestion does not come as a surprise. For the purpose of comparison, San Francisco mid-rise neighbourhoods have a density of some 74 units per hectare (30 units per acre); almost 40 per cent of its population uses public transport, and more than 10 per cent walk or ride a bicycle.

The street layout also reflects the increasing privatization of space in metropolitan areas. The physical layout of recent neighbourhoods strongly discourages people from venturing into neighbourhoods other than their own. There are few points of entry in the contemporary neighbourhood which, in an increasing number of cases, is itself fenced and gated. While the restricted entry points, or the gates and guards, help reduce traffic within the neighbourhood, they also support the further segregation of districts by socio-economic class. Curiously, however, these restricted streets continue to be legally turned over to public authorities (cities or counties) which, in turn, maintain them.

Parks and open space

Until the Second World War, public neighbourhood parks prevailed as the principal form of common open space. They were purchased by towns or donated by private landowners along with land for schools, and usually remained within a ten-minute walking distance from the furthest neighbours. The traditional park occupied between one and four blocks, and the school between one and two blocks. In the 1950s, however, parks and schools were further removed from the neighbourhoods as people were expected to drive to them. Children were increasingly bussed to their schools, and school yards no longer acted as neighbourhood open spaces. The PUD introduced yet another element of the residential subdivision: the private park for the use of residents only. In the garden-apartment complex, the common green became a pretty, but unusable, strip of land between buildings.

CONCLUSIONS

This work suggests lessons for both the management of future land-scapes and research required to improve the understanding of these landscapes.

Morphological study of the North American city is, at this point at least, less related to the issues of historicity with which European cities are struggling, than to issues of dysfunction: the desire for privacy and for segregation by class and land use has led to forms that are no longer congruent with the simultaneous need for high mobility. Historicity is an issue in so far as the forms produced in the early part of the century appear to provide better mobility than the more recent forms (Downs 1989). Some of the social and aesthetic aspects of historicity that European communities are facing may soon also become relevant to early-twentieth-century North American developments. Because these developments have remained relatively free of physical transformations, thanks to the strong policies encouraging suburban expansion, many of them retain valuable traces of the short history of their cities.

Morphological study also suggests that the word 'suburb' may no longer be appropriate for the emerging forms, especially since most so-called suburbs are in fact bona fide incorporated towns. The term 'suburb' was originally used to describe the separate areas abutting a town or city. In its common usage today, however, the word only seems to capture the special character of spread-out, sometimes green, low-density, commercial and office areas, along with usually class-segregated residential districts. Technically, the term is all the more misleading because the suburbs of yesterday are the cities of today. Indeed many of today's originally suburban cities already 'feel' urban. Though it may be too early to relegate the word 'suburb' to the field of urban history, it may be wise to stop using it when dealing with what actually is the twentieth-century North American city (Jackson 1985; Fishman 1987).

More specifically, the study of residential forms and their evolution over the course of the twentieth century provides several lessons regarding the management of residential landscapes. First, the history of the single-family house for the middle class, however brief it has been, shows that the narrow form is a most advantageous solution to providing economical housing. Indeed the wide-and-shallow house appears to be an evolutionary aberration stemming from access to artificially cheap land, an over-optimistic view of the travel range of the private automobile, and an under-estimation of the growth in the number of households that has been characteristic of the latter part of this century. Today, the costs of land and infrastructure have led developers to return instinctively to the narrow house. In its contemporary form, this narrow house is not as deep as its 1920s distant cousin and hence does not suffer from lack of internal light. Also, the negative aspects of the associated narrow side yards are reduced in the common use of semi-detached arrangements.

Secondly, changes in land-development practices have affected the choice in house types and styles now available to homeowners. Increased control by developers over the construction of houses has reduced the number of different house types constructed to a few 'model types'. Moreover, the differences in the siting of houses on their lots and in the interior layout of houses, which were numerous until the 1920s, are relatively few in recent developments. Architectural styles are especially limited in contemporary neighbourhoods, because developers prefer to apply a unified aesthetic which they consider to be a better community image than the practices of the past. Further, the widespread use of covenants to control the outward appearance of new communities promises to stifle adaptive responses to future change. Covenants not only prohibit residents from changing the exterior form or colour of their house, but they also regulate the design and use of the garden, especially the front yard. Given the nature of such covenants, the issue is whether the landscapes of communities will be able to mature along with their residents and be adapted to new needs and tastes. It seems reasonable to subject future covenants to a statute of limitations, which after ten years, for example, would require communities to revise their goals and priorities.

Thirdly, the common practice of organizing new communities around loop roads which discourage through traffic and reduce the number of accidents should be re-evaluated in response to the resulting traffic congestion. Through-streets must be provided at intervals shorter than the quarter-mile (402-metre) grid to increase the level of choice that drivers, cyclists and pedestrians have in circulating from neighbourhood to neighbourhood. The number of loop roads must be reduced because they are *de facto* private roads which only serve adjacent residents. Public authorities need to devise detailed street plans for all areas of town which take into consideration expected traffic volumes, but which also mitigate the likely increased levels of traffic in neighbourhoods – for example, by reducing automobile speed and increasing the number of footpaths and bicycle trails. Streets that are not through-streets should not be subjected to local or state standards.

Last, perhaps the single most important issue to be addressed in managing the residential landscape is density of development and opportunities to gain access to services by means other than the automobile. Contrary to popular belief, the density of contemporary development is intrinsically not a hindrance to safe and efficient transportation. Community plans used in the last two decades typically yield densities that are sufficient to support an adequate public transport system. Further, the densities of the many apartment complexes being built today approximate those needed to facilitate a pedestrian environment (indeed many contemporary communities now require the construction of elaborate pavement systems near such communities). However, on one hand, the limited number of through-streets keeps the web of public transportation too coarse for convenient pedestrian access; on the other hand, the distances between schools, shops, places of employment and residential areas remain prohibitive for comfortable pedestrian travel. Also, zoning practices have ignored the benefits of enforcing walking distances between different land uses, instead favouring the needs of landowners by providing an inordinately high supply of commercial land contained in large parcels. Hence it is not unusual for even the most basic commercial facilities to be dispersed broadly, five or six kilometres apart. PUDs, which could reverse this trend and include public and commercial amenities within walking distance of the houses, are generally single-use residential districts. It seems obvious that planning authorities should now focus on criteria for *proximity* between places of residence, commerce and education, rather than on density, to relieve automobile traffic congestion and to make other means of transport acceptable.

The contents of this chapter also suggest the need for basic research on contemporary forms that goes beyond immediate concerns for landscape management. First, the typologies that are proposed as a theoretical framework, constitute *de facto* hypotheses about significant periods of recent residential development and their morphological characteristics. Clearly, further systematic morphological analyses of particular suburbs are needed to verify these hypotheses. The typologies may not be inclusive. They may benefit from further work in developing subtypes to permit comparisons between morphological characteristics in space and time (synchronically and diachronically). Regional differences can be expected which will point to the relative popularity of different types as well as to reasons for diffusion and adoption of types.

The actual origins of common building traditions and the diffusion of ideas regarding residential development demand the historian's attention.

The work described here shows that there is a considerable lead-time between the articulation of a design idea and its application in a substantial, repetitive fashion. How long is this lead-time, and to what can it be attributed? Similarly, studies should be made of the impact of planning tools on actual development practices. Legislation was designed to encourage PUDs in the early 1960s, yet they only became prevalent in the 1970s.

At the level of the house, J. B. Jackson (1980) set the stage when he talked about the evolution of the garage and the lawn as important socio-architectural elements of the landscape. He and others have inspired many students whose theses are now documenting aspects of the evolution of late-twentieth-century residential forms. More work is needed, for example, on the changes made in typical house plans, the relative openness of the plan, the functional attributes of the rooms, the actual use of the rooms, and the location of kitchens, garages and so on. Similarly, the significant alterations made to private gardens, and the provision of private, but shared, open space, are trends that pose interesting questions as to the new urban dweller's relationship to nature. Further study of these contemporary forms will be essential for the quality of our future habitat.

ACKNOWLEDGEMENTS

The research for this chapter was supported by an exchange fellowship in 1989 with the BOOM Gruppen, KTH, Stockholm (funded in part by the Swedish Council for Building Research), and by the National Endowment for the Arts, a Federal Agency.

NOTES

1   Dimensions of lots, houses, street widths and blocks are given in feet and inches, as they were originally measured, to respect the simplicity of the geometries used and to facilitate the reader's understanding of these geometries. Metric equivalents are then provided in parentheses. They are rounded off to the first decimal point.
2   Chains were the primary means of measurement, where one chain equals 66 feet.
3   Each section of the Land Survey is one mile square, or 5,280 feet square, or 80 chains square.
4   A 10-acre parcel equals 10 chains square.

REFERENCES

Burchell, R. W. (1972) *Planned Unit Development: New Communities, American Style*, New Brunswick, NJ: Center for Urban Policy Research, Rutgers University.

Caniggia, G. (1983) 'Dialettica tra tipo e tessuto nei rapporti preesistenza-attualità, formazione-mutazione, sincronia-diacronia', Extracts from *Studi e documenti di archittetura* 11 (June).

Conzen, M. P. (1980) 'The morphology of nineteenth-century cities in the United States', in W. Borah, J. Hardoy, and G. Stelter (eds) *Urbanization in the Americas: The Background in Comparative Perspective*, Ottawa: National Museum of Man.

Conzen, M. R. G. (1968) 'The use of town plans in the study of urban history', in H. J. Dyos (ed.) *The Study of Urban History*, New York: St Martin's Press.

Downs, A. (1989) 'The need for a new vision for the development of large US metropolitan areas', unpublished paper for the Salomon Brothers Inc., August.

Federal Housing Administration (1938) *Planning Profitable Neighborhoods*, Technical Bulletin 7, Washington, DC: Government Printing Office.

Fishman, R. (1987) *Bourgeois Utopias*, New York: Basic Books.

Foley, M. M. (1980) *The American House*, New York: Harper and Row.

Gowans, A. (1964) *Images of American Living*, Philadelphia: J. B. Lippincott.

Handlin, D. P. (1979) *The American Home*, Boston: Little, Brown and Co.

Hayden, D. (1984) *Redesigning the American Dream*, New York: W. W. Norton and Co.

Heasly, A. E. (1986) 'The front yard: Wallingford landscapes, 1937–1985', unpublished Master's thesis, Department of Landscape Architecture, University of Washington.

Historic Seattle Preservation and Development Authority (1975) 'A visual inventory of buildings and urban design resources for Seattle, Washington', Wallingford Survey, Washington: City of Seattle.

Horowitz, C. F. (1983) *The New Garden Apartment, Current Market Realities of an American Housing Form*, New Brunswick, NJ: Center for Urban Policy Research, Rutgers University.

Housing Authority of the City of Seattle (1940) 'Real Property Survey: 1939–40', Volume 1, General Report, Seattle: Housing Authority.

Jackson, J. B. (1980) *The Necessity for Ruins*, Amherst: University of Massachusetts Press.

Jackson, K. (1985) *Crabgrass Frontier*, New York: Columbia University Press.

Lansing J. B., Marans, R. W. and Zehner, R. B. (1970) *Planned Residential Environments*, Ann Arbor: University of Michigan Press, Survey Research Center Institute for Social Research.

McAlester, V. and McAlester, L. (1984) *A Field Guide to American Houses*, New York: Alfred A. Knopf.

Moudon, A. V. (1986a) *Built for Change*, Cambridge, MA: MIT Press.

Moudon, A. V. (1986b) 'Platting versus planning, housing at the household scale', *Landscape* 29, 1: 30–8.

Moudon, A. V. (forthcoming) *City Building*.

Newman, P. and Kenworthy, J. (1989) *Cities and Automobile Dependence*,

Brookfield, VT: Gower Technical.

Panerai, P., Depaule, J.-C., Demorgon, M. and Veyrenche, M. (1980) *Élements d'analyse urbaine*, Brussels: Éditions Archives d'Architecture Moderne.

Perin, C. (1977) *Everything in Its Place: Social Order and Land Use in America*, Princeton: Princeton University Press.

Puget Sound Council of Governments (1990) *Vision 20/20*, Draft Environmental Impact Statement, Seattle: Puget Sound Council of Governments.

Relph, E. (1987) *The Modern Urban Landscape*, Baltimore: Johns Hopkins University Press.

Seamon, D. and Mugerauer, R. (eds) (1985) *Dwelling, Place and Environment*, Dordrecht: Martinus Nijhof.

Stein, C., (1971) *Toward New Towns in America*, Cambridge, MA: MIT Press. MIT Press.

Stern, R. A. M. (1981) *The Anglo-American Suburb*, Architectural Design Profile, London: Academy Editions.

Stilgoe, J. R. (1988) *Borderland*, New Haven, CT: Yale University Press.

Tunnard, C. and Pushkarev, B. (1981) *Man-Made America, Chaos or Control?* New York: Harmony Books.

Weiss, M. A. (1987) *The Rise of Community Builders*, New York: Columbia University Press.

Whitehand, J. W. R. (ed.) (1981) *The Urban Landscape: Historical Development and Management: Papers by M. R. G. Conzen*, Institute of British Geographers, Special Publication 13, London: Academic Press.

Whitehand, J. W. R. (1987) *The Changing Face of Cities*, Institute of British Geographers, Special Publication 21, Oxford: Basil Blackwell.

Wright, G. (1981) *Building the Dream*, New York: Pantheon Books.

Wright, H. (1935) *Rehousing Urban America*, New York: Columbia University Press.

8

THE PACKAGED LANDSCAPES OF POST-SUBURBAN AMERICA

Paul L. Knox

INTRODUCTION

The high tide of the most recent real-estate and development boom, of 1984–9, has left American cities with some remarkable new landscapes. Private, master-planned communities have appeared – or at least begun to appear – around every large metropolitan area, creating a series of 'artful fragments'[1] that seem likely to prefigure the post-suburban form of the *fin de millénium* metropolis. Unlike their distant antecedents in the Garden City and New Town movements, their provenance is almost entirely from within the private sector, their objectives being concerned less with planning and urban design as solutions to problems of urbanization than as solutions to the problem of securing profitable new niches within the urban development industry. At the same time, they are radically different in scale, layout and composition from the residential subdivisions that have characterized the past forty years or more of metropolitan decentralization. Above all, they are distinctive because of the extent to which they are *packaged*. The hallmarks of private master-planned communities are their packages of amenities and their packaging within a unified design framework.

Private master-planned communities are an important component of an emergent new urban geography, a 'post-suburban' form that stems from the development of the service economy and the materialism of the service class (Kling *et al.* 1991). Post-suburban America is fragmented and multi-nodal, with mixed densities and unexpected juxtapositions of forms and functions. The overall effect has been described by Pierce Lewis as constituting the 'galactic metropolis', where 'the residential subdivision, the shopping centers, the industrial parks seem to float in space; seen together, they resemble a galaxy of stars and planets, held together by mutual gravitational attraction, but with large empty spaces between clusters' (Lewis 1983: 35). There is another aspect of Lewis's analogy that is particularly appropriate. It is simply that much new urban development – commercial and industrial, as well as residential – stands out within the overall built environment because of its glitter, its artfulness and its star quality. As such, private master-planned communities are also part of the 'post-modern turn' that has been inscribed in the built environment not only through architecture but also, to borrow Pevsner's phrase, through mere building. In this context, private, master-planned communities are clearly part of a post-modernism of reaction, a swing of fashion away from the asceticism of modernism (as contrasted with a post-modernism of resistance, which seeks to deconstruct modernism and change its underlying philosophies; see Foster 1985). The packaging that is so characteristic of private master-planned communities can thus be seen as a parallel to the packaged nostalgia, refinement, heritage and sophistication found in the gallerias, shopping malls, office villages, mixed-use developments and festival settings of the post-modern city (Knox 1991). They are a product of stagecraft, proscenia for the enactment of consumption-oriented lifestyles. As such, they are also part of the socio-cultural polarization of the contemporary city, spaces in which estrangement along lines of class and race is heightened even further than in traditional suburbs.

This chapter attempts to set private master-planned communities within the broader sweep of urban change, viewing them as part of the social production of the built environment. First, their antecedents are outlined and their morphological and design attributes established. The extent and character of private master-planned communities are then described with reference to some specific examples. The remainder of the chapter explores the reasons for the emergence of such landscapes, looking in particular at the imperatives of the development industry, of urban planners and designers, and of the new bourgeoisie from whom developers seek customers for their packaged settings.

NEW TOWNS AND NEO-TRADITIONAL PLANNING

The commercial experience and planning concepts that underpin the morphology and design of private master-planned communities derive from a variety of precedents.[2] The US government's green-belt towns of the 1920s and 1930s (Greenbelt, Maryland, Greendale, Wisconsin, and Greenhills, Ohio), together with Radburn, New Jersey, introduced a number of important morphological elements, including clustered housing facing parks and greenways, and town centres with a school and a mix of shops, offices and recreational facilities. The land boom in Florida in the 1920s, meanwhile, produced a number of glitzy 'master suburbs' (including Coral Gables and Boca Raton) with international themes and a strong urban flavour (with pavements, tree-lined boulevards and rectilinear street grids). After the Second World War, some planned Florida suburbs (including Cape Coral, Lehigh Acres and Port Charlotte) that had been established as future retirement settlements quickly evolved into multifunctional communities as young families were attracted by their low housing costs and short commuting distances.

This commercial success, together with the advent of planned unit development (PUD) zoning (whereby building density is aggregated and calculated on a project-wide basis, allowing the clustering of buildings to create open spaces or to preserve attractive site features) in the 1960s, consequently prompted a number of important ventures. In California, planned communities were designed for a broad spectrum of the middle classes, with developers subdividing large tracts of land (such as the Irvine Ranch that became the site of Irvine,[3] the Newhall Ranch, which became Valencia, and the O'Neill Ranch, which became Mission Viejo) into distinctive neighbourhoods with their own housing styles and price ranges and their own recreational amenities. It was the latter – swimming pools, multiple golf and country clubs, urban parks and sports centres – that were the most innovative and successful features of these communities. On the east coast, Columbia, Maryland, and Reston, Virginia, became the showcase examples of American planned communities. In contrast to the Californian communities, they were designed for a rather narrow segment of the established middle classes, and the communities themselves were planned to address aesthetic rather than recreational impulses. Both Columbia and Reston were organized into 'villages' of 10,000 to 15,000 people, with each village having a mix of single- and multi-family housing and its own communal centre, school and recreational facilities. Meadows and woods separated villages from one another and formed the framework for a system of corridors of natural open space followed by walkways and bicycle paths; a large town centre, meanwhile, provided an overall focal point (Breckenfeld 1971; Christensen 1986). Yet, while widely acclaimed for their planning innovations, both Reston and Columbia have had a chequered financial record, only stabilizing as they matured and were completed during the development boom of the late 1980s. The thirteen federally assisted planned communities inspired by Columbia and

Reston in the 1970s were a showcase initiative of the US Department of Housing and Urban Development, but they soon experienced even more serious financial difficulties. As Sternlieb put it, 'the planning drawing-board was filled with aesthetic delight, disastrously coupled with a divine ignorance of cash flow and marketability' (Sternlieb 1987: 22). Planned communities, it seemed at the time, were not to have a significant place in American urban development.

The private master-planned communities that proliferated during the real-estate and development boom of the late 1980s have clearly benefited both from the marketing lessons and planning experience of these earlier ventures. Yet they are more than mere rearrangements of old formulae. They represent a new kind of element in the built environment: thoroughly marketed, closely negotiated and carefully packaged. The concept that has been deployed most widely in marketing, negotiating and packaging these new 'product lines' (as the development industry refers to them) is 'neo-traditional' planning. For potential home-buyers, neo-traditional planning is represented as offering human scale, comfortingly familiar architecture, and a friendly and convenient Main Street. For the civic authorities, with whom the developers must negotiate zoning ordinances and development permits, neo-traditional planning is represented as a solution to traffic congestion, with a mix of office and retailing employment to mitigate inter-suburban travel, pedestrian 'pockets' or villages with built-in transit stops to take as much as possible of the remaining traffic off the roads, pedestrian-oriented shopping and recreation areas, and gridded street patterns that avoid bottlenecks at converging feeder roads. For both home-buyers and civic authorities, neo-traditional planning is represented as a means of achieving distinctive, aesthetically pleasing and high-quality settings that bring packages of amenities such as golf courses, tennis courts and equestrian centres at no direct cost to the taxpayer. On a more abstract level, neo-traditionally planned areas are represented as fulfilling Americans' proclivity for pastoral settings. For Americans, Marx contends,

> regenerative power is located in the natural terrain: access to undefiled, bountiful, sublime Nature is what accounts for the virtue and special good fortune of Americans. It enables them to design a community in the image of a garden, an ideal fusion of nature with art. The landscape thus becomes the symbolic repository of value of all kinds – economic, political, aesthetic, religious
>
> (Marx 1967: 12)

According to Pearson, writing in the trade magazine *The Builder* in 1990, there are six major dimensions to neo-traditional, or 'village', planning.

1   *A mixed-use core*   'Mixing retail, offices and multifamily dwellings (usually condos and apartments above stores) in the town center provides the village with a heterogeneity that contrasts with most PUDs' uniformity of uses and building types. This has benefits not only in terms of convenience but also, supporters suggest, in terms of social integration' (Pearson 1990: 296).

2   *Employment and civic centres*   'Most village developments include a significant amount of office space that could, at least theoretically, provide places of business for most residents. They also provide civic facilities so residents need not always leave the community for recreational and cultural events' (Pearson 1990: 297).

3   *A sense of community*   'To give projects a sense of community – an identity – planners of new villages sometimes make up fictitious histories for them. . . . Another way of giving a development an identity is to reserve prominent sites at the end of major avenues or on hills for landmarks such as historic monuments, public buildings, and churches' (Pearson 1990: 298).

4   *Street life*   'Mixing retail with residential is not enough. . . . Streets must be scaled for pedestrian use, outdoor furniture provided for relaxing, and on-street parking laid out for vehicular access' (Pearson 1990: 298).

5   *Connections*   'Instead of isolating neighborhoods by income or product type, planners of villages try to connect them to each other and to the central business district. This strategy involves a trade-off that developers of PUDs usually don't like: sacrificing privacy for community' (Pearson 1990: 299).

6   *Tradition*   'Although there is no reason [why] village planning principles couldn't be used with modern architecture, all of the villages now being planned employ traditional styles of design. . . . Whether [they] will be able to recapture the charm of places such as Charleston, SC, and Annapolis, MD, is open to debate. What is clear, though, is that a large part of the market is looking for houses that conjure up images of an earlier time' (Pearson 1990: 299–300).

PACKAGED LANDSCAPES

Much of the impetus for neo-traditional planning is attributed in professional magazines to the financial success and critical acclaim of one rather small and, in many respects, atypical project. Seaside, Florida, is a high-income resort 'town' of just 80 acres whose master plan calls for 350 houses and 100–200 apartments (Mohney 1988). It is a product of painstaking research by Andres Duany and Elizabeth Plater-Zyberk into the archetypal American small town, inspired by turn-of-the-century concepts of civic art, *beaux-arts* principles of classical composition, Raymond Unwin's text on *Town Planning in Practice* (Unwin 1909), and Leon Krier's recent polemic on urban design (Krier 1985). Their objective was to write a code for the new town that would re-create an orderly settlement with axes and vistas terminating in identifiable landmarks, and with carefully arranged 'street pictures' with well-proportioned buildings and open spaces functioning as 'public rooms'. The site plan for Seaside features a geometric core surrounded by a grid street pattern that maximizes the number of streets with an ocean vista. The core contains a retail centre conceived as the 'downtown' and a conference facility interpreted as the 'town hall'. The town's code requires that a minimum percentage of the frontage of each lot is built on, with picket fences required on yard frontages in order to define street lines. Porches and arcades of specific dimensions are required for residential and commercial buildings respectively, while outbuildings are encouraged in order to create a 'secondary level of urbanism'.

If Seaside can be taken as the prototype, private master-planned community, then Kentlands can claim, for the present at least, to be the *chef d'oeuvre* of the movement. Kentlands is located on a 352-acre (143-hectare) parcel of farmland 13 miles outside Washington DC. The developer, Joseph Alfandre, having bought the land for $40 million, opted for a packaged, master-planned community rather than the 'cookie cutter' subdivisions that had been his *métier*. Duany and Plater-Zyberk were hired to design the project, and in June 1988 they organized a seven-day 'charette', or group design exercise, that involved architects, builders, traffic consultants, a market researcher, a shopping-mall developer and – critically – politicians and city officials who would have to approve the zoning and road schemes. The charette generated over 500 drawings and culminated in a master plan that established the framework for 1,600 homes (a mix of single-family detached houses, townhouses, apartments, senior housing, condominiums and artist/craft studios), and over 110,000 square metres of office space, all accommo-

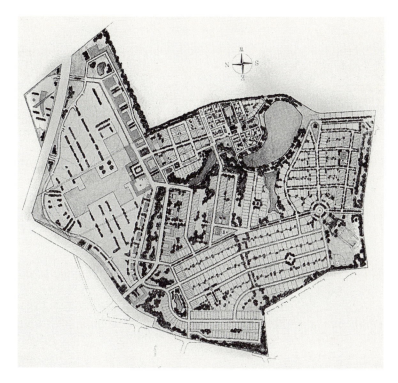

Figure 8.1 Plan of Kentlands (J. Alfandre & Co.).

dated by an entirely new zoning ordinance established by the city of Gaithersburg with the assistance of Alfandre's expertise (figure 8.1). The first of the dwellings was sold in 1990; it is envisaged that the project will reach 'build out' in fifteen to twenty years and that it will take the community another twenty to thirty years to reach maturity. Since Kentlands has, in many ways, become the yardstick for private master-planned communities, it is worth noting the features of its built environment in some detail. Like Seaside, Kentlands is based on a grid street system. It is large enough, however, to warrant several distinct neighbourhoods, and for the grid to be punctuated by a 'Main Street' town centre and public squares. Among the package of features designed to bring authenticity and character to the town are a brick barn, a mill, a firestation, a guest house and a carriage house, all re-created in period style to complement a restored 1852 farmhouse and its outbuildings. These structures and others for which plots have been

213

reserved will house a library, a town meeting hall, churches, an elementary school, a recreation centre and a post office. The package of landscape amenities includes several stands of white oak and spruce pine, four man-made lakes (the largest of which will be equipped with a bandstand/boathouse), a swimming area with bath-houses, a wetlands environment, and an *allée* of cedars and ginkgos. The kitchen garden of the old farmhouse will become the town's 'hidden garden', maintained by a gardener who will also operate a holding nursery to sell plants

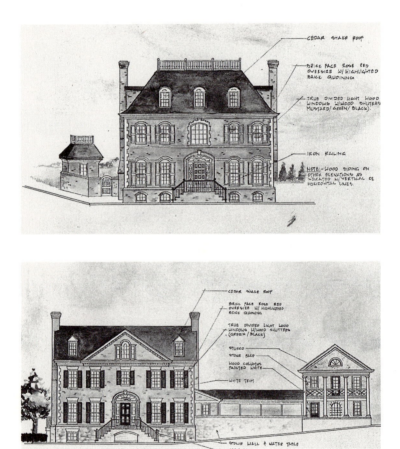

Figure 8.2 Designs for single-family dwellings in Kentlands (J. Alfandre & Co.).

214

deemed appropriate to Kentlands' five neighbourhoods. The dwellings to be built at Kentlands are perhaps the most voluble elements in the rhetoric of post-modern landscapes: allegorical statements that enable their residents to make morally charged statements about themselves and their social relations within the wider community (Duncan 1990). Offered not only by Alfandre and Co. but also by area builders who have bought undeveloped lots at Kentlands, they are overpoweringly traditional – including the tradition of upper-middle-class attempts to mimic the villas, mansions and town homes of genuinely wealthy and socially established families (figure 8.2).

> According to the Urban Land Institute, the essential features of a planned community are a definable boundary; a consistent, but not necessarily uniform, character; overall control during the development process by a single development entity; private ownership of recreational amenities; and enforcement of covenants, conditions and restrictions by a master community association.
>
> (Suchman 1990: 35)

Applying this definition to development projects in the metropolitan fringe of Washington DC in Virginia reveals more than thirty of these 'artful fragments' (though many of them, like Kentlands on the other side of the city, are barely beyond the initial stages of development). Figure 8.3 shows that they are clustered in eastern Loudon County, western Fairfax County and northern Stafford County – along the frontier of the 'galactic metropolis'.

By no means all of them are as large, as carefully researched or as 'neo-traditional' in concept as Kentlands. The overall effect is a collage of private worlds, each entered through substantial portals in the manner of an English country estate, and each announcing itself on large and expensively sculpted and gilded signs with names that draw freely on historic and aristocratic themes, such as Hampton Chase, Manor Gate, Tavistock, Stratford, Century Oak, Exeter and Kingswood. South Wales, it must be admitted, does not fall easily into this characterization; although allowances must be made for American ignorance of geography and naïveté concerning regional class distinctions in Britain. Within these private worlds, the typical package of amenities might include tennis courts, a golf course or driving range, swimming pools, play areas, jogging tracks, an auditorium, exercise rooms, a shopping centre, a daycare centre and a security system operated by electronic key-card systems. Housing is typically dominated by

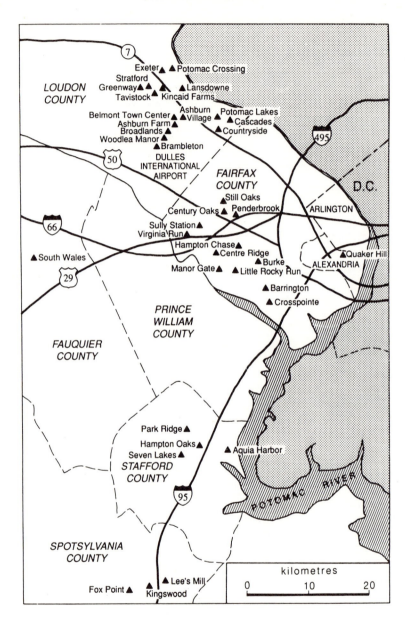

Figure 8.3 The location of private master-planned communities in northern Virginia.

the High Mediocrity of traditional pattern-book motifs – tudor, georgian, colonial, Cape Cod and so on – in overblown dimensions and with an inauthentic abundance of 'architectural detailing' in the form of mouldings, balusters, pilasters, porches and columns. Because the amenity packages have to be achieved at the expense of high densities, the larger single-family houses appear strikingly out of scale, exaggerating the monumental and spectacular quality of the community. The entire ensembles are framed in carefully landscaped settings with generous plantings, expensive fencing and distinctive street signs and mailbox stands. The meticulous tidiness and detailing, combined with the self-conscious affluence of these communities, are reminiscent of a *New Yorker* cartoon depicting a subdivision name plate with a warning appended: 'Traditional Values Strictly Enforced'. Indeed, the po-faced respectability and monumentality of some landscapes evoke a very sombre quality that is only punctuated by busy mail-order delivery vans and a steady stream of joggers and power walkers.

THE STRUCTURES OF BUILDING PROVISION

As Michael Ball (1986) has emphasized, the making – and remaking – of the built environment is a function of sets of time- and place-specific social and economic relations involving a variety of actors, including landowners, investors, financiers, developers, builders, design professionals, construction workers, business leaders, community leaders and, not least, consumers. These sets of relations are what Ball refers to as structures of building provision. They are the animators of morphological change, key elements in the socio-spatial dialectic. The three most important components of the structures of building provision surrounding private master-planned communities are the development industry, the planning profession and the professional service classes for whom they are built. All three must be seen in terms of their broader linkages within the epochal economic and social shifts that have characterized the past fifteen years or so: the emergence of a 'post-Fordist', 'informational' economy (Castells 1989) and the emergence of post-modernity as an important element of the socio-cultural landscape (Harvey 1989a).

The development industry

First of all, it must be recognized that much of the form of post-suburban America is the product of speculative investment in the built environment that has been fuelled by three unusual sets of circumstances. One was the abundance of 'petrodollars' available to holders of oil industry stocks in the wake of the quadrupling of oil prices after the OPEC embargo of 1973. Another closely related, but more fundamental factor was the 'overaccumulation' within the world economy that became apparent soon afterwards. On top of the steady deindustrialization of the labour force, increasing inflation and international monetary instability, and falling rates of profit, the system-shock provided by high energy prices initiated a major round of corporate and spatial restructuring and made it logical for those with capital to invest to turn away from the 'primary circuit' that includes the built environment (Harvey 1978). The third was the deregulation of the savings and loan industry that followed the emergence of the 'New Right' at the end of the 1970s – an ideological shift that was also attributable, in some measure, to the negative economic effects of overaccumulation compounded by the energy crisis. The net result has been a pronounced restlessness in the built environment of US metropolitan areas as a supercharged development industry has inscribed new structures on to the expanding framework of the galactic metropolis.

This, however, helps to explain only the *amount* of development, not its *nature*. In order to understand the latter better, including the appearance of private master-planned communities, increased levels of investment in the built environment must be seen within the context of the restructuring of the corporate world – and, indeed, of the world economy – in the aftermath of the overaccumulation crisis. Like many other industries, the development industry in the US has experienced a significant amount of concentration and centralization, both of which have been reflected in the kinds of products most appropriate to corporate goals and capacities. Concentration involves the elimination of smaller, less profitable firms, partly through competition and partly through mergers and takeovers. Of the housebuilding firms in the US, of which there are more than 100,000, the market share of the top 100 had reached 15 per cent by the end of the 1980s. Centralization involves the merging of the resultant large enterprises from different spheres of economic activity to form large, diversified conglomerate corporations. Among the larger corporations now involved heavily in real estate and development are the likes of Trammel Crow, Maguire Thomas, JMB Realty, VMS Realty, Olympia and York, Prudential Insurance, Equitable and Life, and Metropolitan Life.

As a result of this concentration and centralization, the *scale* of investment projects has increased strikingly. The involvement of conglomerate corporations and large financial institutions has made it possible to put together large-scale, highly visible, high-risk but potentially highly profitable projects. In this context, private master-planned communities can be seen as a means by which large developers are able to create sharply delimited residential submarkets and thus extract monopoly rent (see King 1989 for a discussion of this concept). At the same time, competition between developers has resulted in an escalation of the range of amenities and the distinctiveness of the packaging, thus requiring even larger projects. The costs of farmers' markets, exercise rooms and wetlands environments can only be carried by developers who can afford to invest 100–200 million dollars over about ten years before seeing any return. Because the success of private master-planned communities (in commercial terms) depends on a race between rising carrying costs (mainly interest) and rising land values, their character and their morphogenetic sequencing are tied to the tensions and contradictions of real-estate financing. In order to foster increases in land values, developers are impelled (i) to establish elaborate and comprehensive master plans that do not allow opportunities for a dilution of the quality of the enterprise, (ii) to secure all development approvals at an early stage, (iii) to provide infrastructure, amenities, 'signature' landscaping and upscale homes as soon as possible. In order to keep carrying costs down, meanwhile, some of the most expensive amenities have to be deferred as long as possible, relying on skilful marketing to reassure buyers of the imminence of their appearance. Other means of keeping carrying costs down include seeking 'patient' capital (thereby allowing greater flexibility to develop and sell property when the market is expensive and to pull back when it is in recession), shifting the costs of infrastructure provision on to the public sector through public–private partnerships or through the creation of special districts (with authority under state laws to issue tax-exempt bonds and to collect property taxes or fees to repay the bonds), and generating other sources of income by creating sidelines such as real-estate brokerage, golf-course management, cable TV or sewer utilities (Ewing 1990b).

These strategies underline another broad change in the organization of the industry: the need for flexibility and the importance of niche marketing. Though less pronounced in the development industry than in certain areas of manufacturing, there has been a shift away from a reliance on economies of scale through the pursuit of mass production and mass consumption (the so-called 'Fordist' approach to production, epitomized in the development industry by Levittown) towards 'lean' and 'flexible' production systems and product-differentiation strategies that reach the most potentially profitable niches within the market. As Harvey puts it,

> the frenetic pursuit of the consumption dollars of the affluent has led to a much stronger emphasis on product differentiation.... Producers have, as a consequence, begun to explore the realms of differentiated tastes and aesthetic preferences in ways that were not so necessary under a Fordist regime of standardised accumulation through mass production.
>
> (Harvey 1987: 273–4)

In this context, private master-planned communities represent one of a number of new product lines that includes retail gallerias, specialized malls, office villages, 'flexspace',[4] mixed-use complexes and spectacular waterfront and festival marketplace settings (Knox 1991). Private master-planned communities not only provide a product with potentially high profit margins but do so within a framework that permits developers more flexibility in design and product type and enables them to respond quickly to changing market demand (Suchman 1990: 38). The packaging inherent in private master-planned communities (and in most of these other new product lines) provides a distinct marketing advantage over traditional forms of residential development, enabling advertising imagery to draw on the totemism of stylish materialism as the basis for exclusive lifestyles. Long Signature Homes, for example (available in several master-planned communities in northern Virginia, including Virginia Run, Sully Station and Cascades), are advertised as offering a 'picture-perfect lifestyle', while Woodlea Manor is described as a place 'Where Style of Living Matters'. Advertising copy for homes in private master-planned communities is suffused with references to luxury, privilege, prestige, tradition, authenticity, detail, richness, crafting and styling.

Planners

Exhausted by the unrealizable utopianism of the Brave New World of the 1950s, 1960s and 1970s (Brooks 1988) and undermined by the shift in civic culture towards entrepreneurialism (Harvey 1989b), city planning has lost its way. Estranged from planning theory and divorced from any broad sense of public interest, it has become increasingly geared to the needs of producers and the wants of consumers; less concerned with overarching notions of rationality or criteria of public good (Friedmann 1987; Sternlieb 1987; Cuff 1989; Beauregard 1990). The outcome has been a disorganized approach that has fostered a collage of highly differentiated spaces and artful fragments. As Christine Boyer observes, 'fragmented elements of the city whole are planned or redeveloped as autonomous elements, with little relationship to the whole and with direct concern only for adjacent elements' (Boyer 1987: 6). Much of this activity, moreover, is negotiated through public/private partnerships, which have emerged as the foundation of the new, entre- preneurial civic culture. These partnerships, wherein public resources and/or legal powers are joined with private interests, have engendered a speculative and piecemeal approach to the management of urban growth, with attention being focused on localized, high-profile projects (Harvey 1989b). As a result, 'planning has become entrepreneurial and planners have become dealmakers rather than regulators' (Beauregard 1989: 388). The deals that they make include concessions over the nature, scale and location of development or the financing of infrastruc- tural elements in return for donations of land for public schools or parks, for the inclusion of daycare facilities, or for streetscape improve- ments (Alterman 1990; Lassar 1989). For developers of private master- planned communities, the most important concessions concern zoning ordinances. In order to facilitate the kind of projects involved, planners have had to abandon single-purpose zoning – once the cornerstone of planning practice. Mixed-use zoning, combined with PUD zoning and detailed ordinances keyed into master-plan design concepts, is a pre- requisite, allowing developers to exploit economies of scale while incor- porating a comprehensive package of ingredients, all within a flexible but predictable regulatory framework.

While public-sector planning has been left to wring its hands over the loss of professional power and vision, new 'planners' such as James Rouse (a mortgage financier and developer) and Andres Duany (an architect) have emerged from private practice as evangelists for the 'Brave Old World' of neo-traditional planning. Duany, for example, has been lauded by *Architecture* magazine for 'breaking the code ... changing the way America is planned' (April 1990 issue). The business magazine *Warfield's* featured Duany on its April 1990 cover under the headline RESCUING SUBURBIA, with the story inside announcing 'archi- tect Andres Duany and builder Joe Alfandre team up to save the suburbs from themselves' (p. 60). The same month's issue of *Metropolitan Home* (the handbook of post-modernity in American domestic culture) also featured Duany, concluding that 'when we look at the work of Duany/Plater-Zyberk, we can confidently say we've seen the future and it looks livable' (p. 123). *People* magazine, meanwhile, ran a story in its 1990 Spring Extra issue on Duany and Plater-Zyberk under the heading BANISHING COOKIE-CUTTER 'BURBS, portraying them as 'two of America's most forward-looking town planners' who are 'putting front porches, strollable streets and old-fashioned neighborliness back on the map of suburban America' (p. 72).

This kind of attention is undoubtedly the product of an efficient publicity office; but in a broader context it is a consequence of developers' need for 'name' designers who can secure visibility and distinction within a competitive market. It also raises some interesting issues concerning the social construction of landscapes and spaces and the roles of the media in interpreting and imagining urban change (Knox 1992). At the same time, the vehicles of the publicity – glossy magazines – point to the importance of the principal audience/market for packaged landscapes: the affluent professional classes who constitute the 'new bourgeoisie'.

The new bourgeoisie and material culture

While there has long been a demand among upper-middle-class households for up-market housing in distinctive suburban subdivisions, the proliferation of 'packaged' landscapes in general and of private master-planned communities in particular must be attributed at least in part to the consolidation of an affluent and materialistic class fraction that has sponsored – and been left legitimated by – a lifestyle of stylish materialism. Bourdieu (1984) suggests that this class fraction is drawn from the new bourgeoisie and the new petite bourgeoisie. Following Bourdieu (1984), the established bourgeoisie consists principally of intellectuals and industrial and commercial employers, whereas the new bourgeoisie consists of members of the professions, public administrators, scientists and private-sector executives, especially those involved in non-material products – economists, financial analysts, management consultants, personnel experts, designers, marketing experts, purchasers and so on. And, whereas the established petite bourgeoisie consists of artisans, small shopkeepers, office workers and the like, the new petite bourgeoisie consists of junior commercial executives, engineers and skilled technicians, medical and social-service personnel and, in particular, people directly involved in cultural production – authors, editors, radio and TV producers and presenters, magazine journalists and so on. Members of the baby-boom counter-culture generation, they have shifted the focus of their lifestyle from collectivist approaches to the exploration of freedom and self-realization to introverted approaches characterized by materialism, narcissism and hedonism. The reasons for this shift are bound up in the effects of increased competition in housing and labour markets, the struggle for cultural hegemony, the decreased effectiveness of mass-produced consumer goods as status symbols, the incapacity of the American middle classes for ironic reflection, and the overall shift towards a post-modern sensibility (Bourdieu 1984; Baudrillard 1989; Ehrenreich 1989; Knox 1992). The net effect has been the emergence of an affluent class fraction preoccupied with stylish materialism. A central issue for its members is how, in the scramble for social distinction, to avoid being caught in a compromising position with *déclassé* people, objects and places. In this context, the private master-planned community comes into its own as a stylish setting for stylish materialism, a vivarium for narcissistic and hedonistic lifestyles, and a physical framework for the development of a distinctive collective *habitus.*[5]

CONCLUSION

Private master-planned communities have been presented as a characteristic new element of the post-suburban form of the *fin de millénium* American metropolis, a product of intersecting shifts in the dynamics of the urban development industry, of the design professions, and of material culture. As such, their significance lies less in the *chic manqué* of neo-traditional planning than in their role within the socio-spatial dialectic. As outcomes of social, economic and cultural change, they bring to the urban landscape the imprint of post-modernity and flexible accumulation. As mediators of social, economic and cultural change, they remain something of an unknown quantity. Nevertheless, there are several issues of potential significance to the evolution of urban social geographies and, thus, to the continuing evolution of urban physical configurations. It seems likely, for example, that private master-planned communities, because of their scale and their physical separation from the rest of the urban fabric, will act as forcing grounds for stylish materialism and cultural privatization. The result will be a series of communities of affect that exhibit little communality, sharing only a dedication, on the one hand, to the iconolatry of visible wealth and distinction and, on the other, hostility or indifference to the less affluent, less stylish and less fortunate (Ehrenreich 1989). The artful fragment represented by the private master-planned community is thus set to constitute, and reconstitute, part of an increasingly polarized, lifestyle-oriented and spatially segregated social order. The emergent culture fostered by the everyday experience of the sequestered settings of private master-planned communities, meanwhile, carries important implications for other dimensions of urbanization, not least of which will be the politics of collective consumption and, in particular, of educational resources.

ACKNOWLEDGEMENT

Figures 8.1 and 8.2 are reproduced by courtesy of J. Alfandre & Co.

NOTES

1   The term is Christine Boyer's, though she uses it in a more general sense to refer to the outcomes of an urban development process that has become disconnected from the broad visions of modernist urban planning. See Boyer (1987).
2   The following summary is based on the account in Reid Ewing (1990a).
3   See Schiesl (1991) for a detailed account.
4   A term used for single-storey structures with 'designer' frontages, loading docks at the rear, and interior space that can be used for offices, R & D laboratories, storage or manufacture, in almost any ratio.
5   This is a collective perceptual and evaluative schema that derives from its members' everyday experience and operates at a subconscious level, through commonplace daily practices, dress codes, use of language, comportment and patterns of consumption. The result is a distinctive pattern in which 'each dimension of lifestyle symbolizes with others' (Bourdieu 1984: 173). Each class fraction establishes, sustains and extends its *habitus* through the appropriation of 'symbolic capital': luxury goods attesting to the taste and distinction of the owner.

REFERENCES

Alterman, R. (1990) 'Developer obligations for public services in the USA', in Healey, P. and Nabarro, R. (eds) *Land and Property Development in a Changing Context*, Aldershot: Gower.
Ball, M. (1986) 'The built environment and the urban question', *Environment and Planning D: Society and Space* 4: 447–64.
Baudrillard, J. (1989) 'After Utopia' (interview), *New Perspectives Quarterly* 6: 52–4.
Beauregard, R. (1989) 'Between Modernism and Postmodernism: the ambivalent position of U.S. planning', *Environment and Planning D: Society and Space* 7: 381–96.
Beauregard, R. (1990) 'Bringing the city back in', *Journal of the American Planning Association* 56: 210–15.
Bourdieu, P. (1984) *Distinction: A Social Critique of the Judgement of Taste*,

London: Routledge and Kegan Paul.

Boyer, C. (1987) 'The return of the aesthetic to city planning: future theory as a departure from the past', paper presented to the Conference on Planning Theory in the 1990s, Center for Urban Policy Research, New Brunswick, Rutgers University.

Breckenfeld, G. (1971) *Columbia and the New Cities*, New York: Ives Washburn.

Brooks, M. (1988) 'Four critical junctures in the history of the urban planning profession', *Journal of the American Planning Association* 54: 241–8.

Castells, M. (1989) *The Informational City*, Oxford: Blackwell.

Christensen, C. A. (1986) *The American Garden City and the New Towns Movement*, Ann Arbor: UMI Research Press.

Cuff, D. (1989) 'The social production of built form', *Environment and Planning D: Society and Space* 7: 433–48.

Duncan, J. S. (1990) *The City as Text: The Politics of Landscape Interpretation in the Kandyan Kingdom*, Cambridge: Cambridge University Press.

Ehrenreich, B. (1989) *Fear of Falling*, New York: Pantheon.

Ewing, R. (1990a) 'The evolution of new community planning concepts', *Urban Land* 49, 6: 13–17.

Ewing, R. (1990b) 'Financing new communities', *Urban Land* 49, 8: 10–15.

Foster, H. (ed.) (1985) *Postmodern Culture*, London: Pluto Press.

Friedmann, J. (1987) *Planning in the Public Domain*, Princeton, NJ: Princeton University Press.

Harvey, D. W. (1978) 'The urban process under capitalism: a framework for analysis', *International Journal of Urban and Regional Research* 2: 101–31.

Harvey, D. W. (1985) *The Urbanization of Capital*, Oxford: Blackwell.

Harvey, D. W. (1989a) 'From managerialism to entrepreneurialism: the transformation in urban governance in late capitalism', *Geografiska Annaler* 71B: 3–17.

Harvey, D. W. (1989b) *The Condition of Postmodernity*, Oxford: Blackwell.

King, R. J. (1989) 'Capital switching and the role of ground rent: 3. Switching between circuits, switching between submarkets, and social change', *Environment and Planning A* 21: 853–80.

Kling, R. J., Olin, S. and Poster, M. (1991) *Postsuburban California. The Transformation of Orange County since World War II*, Berkeley: University of California Press.

Knox, P. L. (1991) 'The restless urban landscape: economic and sociocultural change and the transformation of metropolitan Washington, DC', *Annals, Association of American Geographers* 81, 2: 181–209.

Knox, P. L. (1992) 'Capital, commodity aesthetics and the built environment', in Knox, P. L. (ed.) *The Restless Urban Landscape*, Englewood Cliffs, NJ: Prentice Hall.

Krier, L. (1985) 'Visionary architecture', *Architecture and Urbanism* 11: 91–100.

Lassar, T. J. (1989) *City Deal Making*, Washington, DC: Urban Land Institute.

Lewis, P. F. (1983) 'The galactic metropolis', in Platt, R. H. and Macinko, G. (eds) *Beyond the Urban Fringe*, Minneapolis: University of Minnesota Press.

Marx, L. (1967) *The Machine in the Garden*, Oxford: Oxford University Press.

Mohney, D. (1988) *Seaside*, Princeton: Princeton University Press.

Pearson, C. (1990) 'The new new towns', *The Builder*, January: 294–301.

Schiesl, M. J. (1991) 'Designing the model community: the Irvine Company and suburban development', in Kling, R. J., Olin, S. and Poster, M. (1991) *Postsuburban California. The Transformation of Orange County since World War II*, Berkeley: University of California Press.

Sternlieb, G. (1987) 'Planning, American style', *Society* 25, 4: 21–3.

Suchman, D. R. (1990) 'Housing and community development', in Schwanke, D. (ed.) *Development Trends 1990*, Washington DC: Urban Land Institute.

Unwin, R. (1909) *Town Planning in Practice: an Introduction to the Art of Designing Cities and Suburbs*, London: T. Fisher Unwin.

9

THE CHANGING SUBURBAN LANDSCAPE IN POST-WAR ENGLAND

J. W. R. Whitehand, P. J. Larkham and A. N. Jones

INTRODUCTION

The recycling of urban land is a major phenomenon in Western cities. Some of the areas being recycled for more intensive development in Great Britain contain some of that country's best housing, often large, architect-designed houses in extensive plots. This reflects a large demand for more, and different, dwellings in many parts of Great Britain. It has been particularly evident in South-East England (Damesick 1986; Simmons 1986; Hamnett 1986). It has apparently been fuelled by a reduction in the average size of households, increases in disposable incomes, and the redistribution of population.

The strict control over new development around many large cities in Great Britain, for example to protect Green Belts and Areas of Outstanding Natural Beauty, is serving to concentrate new developments within existing urban areas. Few recent attempts to produce new settlements have, over the past decade, been successful, as is demonstrated by the fate of recent proposals listed in Amos's directory of new settlement schemes (Amos 1991). In South-East England, some two-thirds of new residential developments are within existing urban areas (Tym and Partners 1987: 6). Bibby and Shepherd (1991) suggest that, for the UK as a whole, 45 per cent of new housing in the 1990s will be 'urban infill'. In the mid-1980s, Goodchild and Munton suggested that smaller urban sites were likely to become of increasing importance in the development process (Goodchild and Munton 1985). This forecast has already been shown to be correct, particularly in those mature residential areas, containing large detached houses on extensive plots, that lie on the fringes of many British towns. These areas have high-quality urban landscapes, high existing use values and, compared with 'greenfield' sites, highly fragmented landownership. Where individual gardens are large, there is little problem in assembling sites of sufficient size to make redevelopment or infill profitable; indeed, during the 1980s, even some large housebuilders became interested in individual plots of under 1 acre (see Whitehand 1989a for an example).

The amount of pressure for such development in South-East England is reflected in a variety of aggregate statistics compiled by local planning authorities (LPAs) for the Department of the Environment (DoE). For example, figure 9.1 shows that a high proportion of planning applications took over eight weeks (the statutory period) to decide in this region in the mid-1980s. The length of time taken to reach a decision is a function of the number of applications, the complexity of the issues raised, and the staff and facilities available for processing applications. There is also in South-East England a relatively high incidence of appeals to the Secretary of State for the Environment against refusals of planning permission.

The considerable variations between the regions of the UK in the pressure for more intensive development of existing residential areas are germane to an understanding of geographical variations in the changing form of urban areas (Pompa 1988; Whitehand 1988; Booth 1989; Whitehand and Larkham 1991a, b). This chapter considers the processes by which building densities are being increased in residential areas in both South-East England, a region of heavy development pressure, and the English Midlands, a region of relatively light development pressure in recent decades. The nature and extent of changes to urban landscapes and the activities responsible for their implementation are examined. The interaction between developer and LPA is of major importance. In many cases this relationship involves conflicts, the nature and consequences of which are examined in detail.

The location of mature residential areas of large detached houses has been documented for South-East England (Whitehand 1967) and for less extensive areas elsewhere in the UK (Cornish 1988; Whitehand and Larkham 1991b). Examination of the complex issues involved in redevelopment of these residential areas requires the detailed investigation of individual plots. A survey of a stratified random sample of eight 25-hectare National Grid squares on London's residential fringe carried out in 1987 suggests that over 50 per cent of the original plots have been affected by some form of change leading to increased densities (Whitehand 1988: figure 2). The large amount of time required to compile the information necessary to elucidate the processes leading to these higher densities severely limits the number of areas and sites that it is practicable to study. It necessitates a case-study approach. Despite the

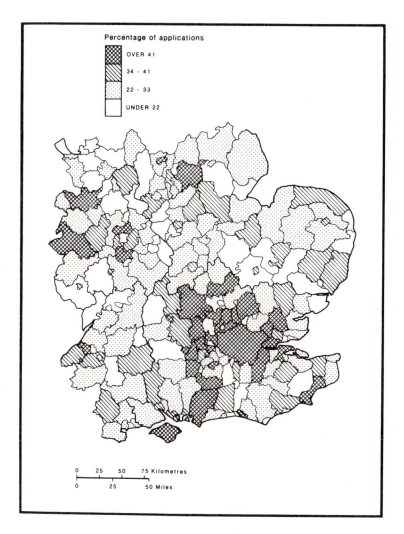

Figure 9.1 Percentage of planning applications decided in over eight weeks in the English Midlands and South-East England, 1983/84–1985/86. Data from Department of the Environment 1987.

problems of such studies, not least those of representativeness, recent research supports the view that they have considerable advantages. These include the ability to retain a holistic and realistic view of actual events, and the possibility of in-depth analysis of a multiplicity of causal links (Punter 1989: 55–6).

Six areas, each comprising a 25-hectare National Grid square, were selected for detailed study. The three squares in South-East England were at Epsom, Northwood and Amersham, and the three squares in the English Midlands were at Tettenhall (Wolverhampton), Gibbet Hill (Coventry) and Barnt Green (Worcestershire). Each of these six study areas was developed as an upper-middle-class commuter area almost entirely between late-Victorian times and the mid-twentieth century. In the 1950s the majority of each square consisted of residential plots of at least 0.15 hectares.

For each study area, plot histories were pieced together from LPA files. All planning applications for new buildings submitted between the beginning of 1960 and mid-1987 (the date when the data collection began) were inspected. Applications submitted before 1960 were included if they were part of a series of related applications the most recent of which was received by the LPA within the study period. Further information was obtained from discussions with local planning officers, site owners, developers and architects. The data were analysed by site. For this purpose a site was defined as a plot or group of contiguous plots that was, at some point in the study period, treated as an entity for the purpose of undertaking development. Nearly 250 sites were examined in detail, and about 1,000 individual planning applications were studied.

The degree of intra-regional variability in the amount and type of change in such areas (Whitehand 1988: 354; Whitehand and Larkham 1991a) suggests that it would be unrealistic to expect these particular areas to be fully representative of mature, low-density residential areas in the two regions. Nevertheless, they do provide a reasonable starting-point for inter-regional comparisons, and collectively suggest some of the characteristics of change and attempted change to the urban landscape that are common to this type of area.

THE AMOUNT OF CHANGE

The amount and type of change in these areas were examined in the field and from the LPA files. The number of sites affected by development proposals was almost identical in the two regions (122 sites in the

Table 9.1 Number of sites that were the subject of planning proposals, and planning gestation periods (in months), in six study areas, *c.* 1960–87.

| | Number of sites | Gestation period | |
		Mean	Median
Amersham	28	58	23
Epsom	45	29	13
Northwood	48	60	14
SOUTH EAST	121	49	17
Barnt Green	44	48	3
Gibbet Hill	22	62	47
Tettenhall	56	24	3
MIDLANDS	122	45	3

Note: One site on which the local-planning-authority decision was pending on 30 June 1988, three sites where the submission date of the initial application is unrecorded and four sites on which the most recent application was withdrawn, date unknown, are excluded

Midlands and 121 in the South East), but there was considerable variation within each region (table 9.1). Areas that have been redeveloped or more intensively developed or, in a few cases, previously undeveloped areas that have been developed for the first time, are shown in black in figure 9.2. Sites on which development had been proposed, but had not been carried out by the end of 1989, are shown stippled. In all the study areas a large amount of land has been the subject of proposals for redevelopment or more intensive development. The extent of this land does not vary much between the Midlands areas and the South-East areas, but there is a marked difference in the density at which the new development has taken place. Whereas in the South-East areas 770 dwellings were constructed, or under construction, during the study period (312 at Epsom, 277 at Northwood and 181 at Amersham), the comparable figure for the Midlands areas was only 195 dwellings (95 at Tettenhall, 29 at Gibbet Hill and 71 at Barnt Green).

The most significant decisions about development tend to be taken before construction begins, and thus the time span, often prolonged, between the selection of a potential development site and the ultimate approval, sometimes disapproval, of a development scheme is of interest. The closest approximation to this time span that may be calculated for a large number of sites is the 'gestation period', defined as the time-lag between initial application and approval of the development

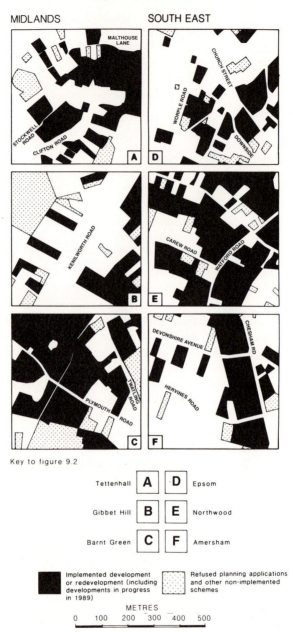

Figure 9.2 Plots subject to attempted or actual development/redevelopment in six study areas, *c.* 1960–87.

that was implemented or, in cases where no implementation took place, the most recent rejection or approval (table 9.1). The median gestation period is higher for the sites in the South East than for those in the Midlands, although there is considerable variation between study areas. The Tettenhall area has a large number of sites that received planning permission relatively quickly, whereas the Gibbet Hill, Amersham and Northwood areas had many sites with protracted planning histories, in several cases having gestation periods of over two decades. The Northwood area, in particular, shows evidence of high development pressure, and there are several sites where a plot created in an initial development has been subdivided or redeveloped and then at a later date subjected to a further development.

THE PHYSICAL CHARACTER OF CHANGE

Types of development

Figure 9.3 is an attempt to depict the main types of development in the study areas. The categories shown combine the physical form of development, or redevelopment, and the way in which that development fits into the existing plot pattern. Where more than one category is represented within a single development, the predominant type (in most cases measured by number of dwellings) is shown.

Flats (including maisonettes) are much more numerous in the South-East areas, there being 26 developments in which flats are the predominant dwelling type in those areas, compared with only two in the Midlands areas. This reflects the greater pressure on land in the South East. The fact that redevelopment was much more common in the areas there (41 developments involving redevelopment, compared with only 16 in the Midlands areas) should be viewed in the same light. In other respects there are no significant regional differences, the striking feature being the variations between and within individual study areas. Probably the most important factors underlying these variations are the plot pattern and street system of the initial development. Thus the deep plots of the Barnt Green area lend themselves to the filling-in of plot tails, access being gained by culs-de-sac from existing roads ('rear infill' in figure 9.3). The wide plots of the Amersham area, in contrast, are more suitable for the lengthways division of each plot and the insertion of an additional house alongside the original house ('frontage-division infill' in figure 9.3). However, neither of these means of accommodating additional dwellings is favoured by highway authorities where the

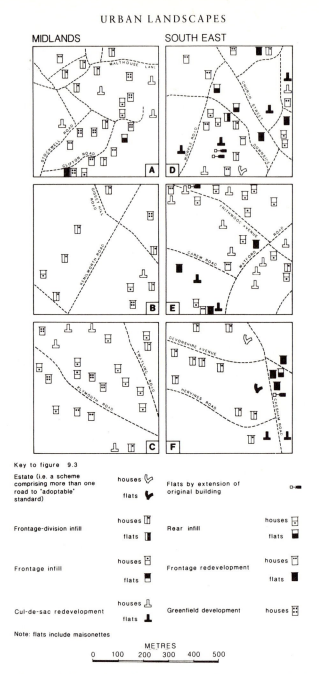

Figure 9.3 Types of development in six study areas, *c.* 1960–87.

frontages of existing plots are on main roads, because of the additional main-road accesses that they entail. Thus various types of redevelopment, rather than infill, have provided the main means of increasing dwelling densities in Chesham Road in the Amersham area and in Watford Road in the Northwood area.

Conversion of ancillary buildings

A feature of several of the study areas was the conversion of ancillary buildings, especially coach-houses and garages of large Victorian villas, and in two cases gate-houses, into free-standing dwellings for separate households. Although this phenomenon is not represented in figure 9.3, which refers solely to new construction, it has contributed to the increasing residential density in several of the study areas. It allows retention of what are, in some cases, visually significant smaller buildings. At least one LPA considered the creation of dwellings out of non-residential buildings to be sufficiently important to insert a policy statement on the subject in its Structure Plan in 1975.

> Subject to the general policy of not normally permitting new houses in open country, consideration will be given to satisfactory proposals for the conversion of suitable non-residential buildings to living accommodation, if by doing so buildings in keeping with the character of the area are thereby restored and retained, provided that there is no over-riding objection to the development.
>
> (Worcestershire County Council 1975)

Spurred by the continuing pressure for such conversions, Bromsgrove District Council adopted detailed guidelines for applicants that were referred to in several proposals in the Barnt Green study area. Of particular relevance to the conservation of the urban landscape are their guidelines that the character and detailing of the original building shall be retained and existing openings utilized wherever possible. New openings shall be kept to a minimum and the style of fenestration shall be appropriate to the building. There shall be no introduction of 'alien' external features such as dormer windows, bow windows, obtrusive chimneys, modern patio doors, etc., unless these are already features of the building (Bromsgrove District Council 1984).

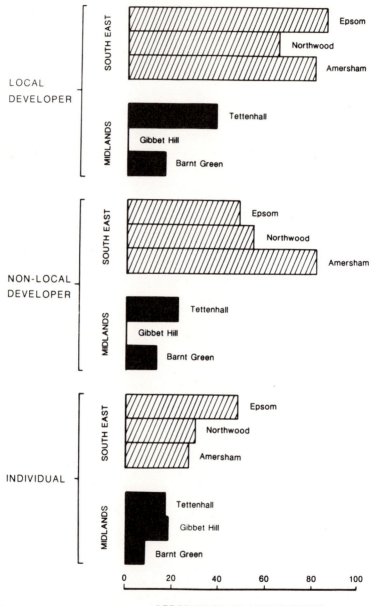

Figure 9.4 Percentage of planning applications, for the construction of new dwellings, entailing demolition of a main building: six study areas, *c.* 1960–87.

Demolition

The demolition of the main building on a plot – the 'plot dominant' – generally involves a drastic change in the urban landscape. A considerable proportion of applications for new dwellings proposed this type of demolition, but the proportion was much higher in the areas in the South East (on average nearly 60 per cent) than in those in the Midlands (on average only about 15 per cent). All types of applicant were much less demolition prone in the Midlands areas. In all areas the proportion of implemented developments entailing demolition tended to be smaller than the proportion of applications entailing demolition. LPAs may have deterred the submission of applications to demolish plot dominants, for example, by policy statements and pre-submission negotiations. For much of the study period, most of the study areas contained designated Conservation Areas, where the demolition of all buildings has been subject to control since the 1974 Town and Country Amenities Act.

By comparison with developers, individuals, most of them local, submitted a low proportion of applications entailing demolition (figure 9.4). This may, in part, reflect the primary concern of many owner-occupiers to establish the principle of more intensive development as a basis for selling land to a developer, usually at a higher price than that of the site in its existing use. A further contributory factor to a lower proportion of demolition applications from individuals was the evident desire of some owner-occupiers either to continue to occupy the existing house, or to occupy a newly created dwelling in the garden. Surprisingly, in both regions non-local developers, who might have been expected to be least imbued with a sense of place, were less demolition prone than were local developers.

CONFLICT AND CHANGE

Conflict is evident at a number of scales in this process of change. At the large scale it is apparent in local politics and, more importantly, national politics, which influence, for instance, alterations to Structure Plans by the Secretary of State for the Environment. For example, the decision by Nicholas Ridley, when Secretary of State, to add 3,000 houses to the number proposed in the Surrey Structure Plan led to considerable local and national political controversy (*Chartered Surveyor Weekly* 1988: 25). At the medium scale there are differences between developers' wishes and adopted policy in the form of local plans, structure plans and, most recently, unitary development plans, and in the definitions of particular areas for planning-policy purposes, such as Conservation Areas or areas of housing restraint. At the micro-scale, conflicts are frequently evident between the developer of a site, the immediate neighbours and LPA case officers. There is little recognition in local plan documents of the attributes that endow areas with their physical character, how such attributes change over time, and how development proposals may alter or reinforce them (Whitehand 1989a, 1990a). But conflict at this smallest scale is fundamental to the resistance to development that many LPAs exercise – a resistance that tends to be popular with voters.

The process of conflict, operating at various scales, and the variety of intended and actual results in the urban landscape, are best illustrated by a series of case studies drawn from the six study areas.

CASE STUDIES

Northwood municipal housing sites

The example of housing sites developed by the local authority in the London Borough of Hillingdon shows conflict at a relatively large scale, including disagreements between Hillingdon and the neighbouring district, conflict within the local authority itself, and the formation of a vociferous local pressure group to resist the proposals. An important part of the context of these developments is the Holland Report (1965), which not only stressed the need for private development by landlords and housing associations to solve the growing urban housing crisis, but also supported a large-scale municipal housing drive. Following the 1972 local-authority elections, the Labour-controlled London Borough of Hillingdon approved an Action Area Plan for development in the Eastbury area of Northwood. Important elements in the plan were the preservation of trees, the preclusion of high-rise development, the encouragement of new buildings that reflected the existing urban landscape, the provision of a primary school and more open space, and the enforcement of an upper residential-density limit of 60 habitable rooms per acre (hrpa) (London Borough of Hillingdon 1973).

To achieve these objectives, the Council designated for municipal housing development a number of sites assembled out of the gardens of existing dwellings. The largest such site lay partly within the Northwood study area (figure 9.5). It appears that the greatest driving force for municipal development in this upper-middle-class area was

238

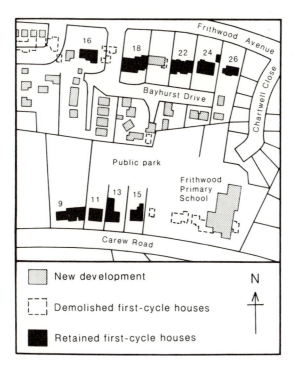

Figure 9.5 Part of municipal housing site, Northwood, redrawn from London Borough of Hillingdon 1973 and field survey.

social restructuring. The design and quality of development were not important considerations. Rapid implementation was important, however, and council applications for the redevelopment of these sites were passed by the planning committee with some speed, much to the concern of local residents and Conservative councillors. These parties claimed that the LPA was acting illegally, was politically corrupt and was, above all, reducing the value of their properties by developing a 'council housing estate' so close to their homes. Twenty years after the event, some of those involved within the LPA admit, unofficially, that they were guilty of bad planning practice, driven by political ideology (Jones 1991: 151, 417). The development was also opposed by the neighbouring local authority, Three Rivers District Council, which was concerned that the process of public inquiry was being by-passed, as this development would decide the future of a sizeable area before proposals for a local plan could be considered. Three Rivers District Council directed that:

> the London Borough of Hillingdon be informed that this council is concerned with the magnitude of the redevelopment proposals with respect to the provision and general availability of essential services and facilities and would prefer to see this matter reconciled in the District Plan for Northwood before significant development proposals are entertained.
>
> (Three Rivers District Council Minutes, 10 September 1975)

The proposal for the largest site involved the compulsory purchase, and the demolition or conversion, of thirty-six large, first-cycle dwellings. That this was a politically motivated proposal is shown by comments by officers of the borough, who were not convinced of the wisdom of the scheme. The Local Plans section noted, before the proposal was formally submitted, that 'this area is better suited to private development of the nature already developed in the post-war period' (planning files, London Borough of Hillingdon). Furthermore, the speed and scale of development, together with the compulsory purchase and major social restructuring of the area, prompted action on the part of local residents. Twenty letters of objection were lodged with the LPA within a week of the planning application being submitted, and a total of 1,849 letters were delivered relating to all such proposals in the study area. According to planning officers, this is by far the largest number of objections received by this LPA for any sequence of applications relating to a single development site.

The physical effects of this development cannot be divorced from the manner in which land acquisition, design and planning took place. The new urban landscapes were developed at dwelling densities over double those previously existing. As the local authority required such a large number of new, smaller dwellings, new types of unit were necessary. 'Linked dwellings', including two-storey terraces, three-storey 'town houses' and apartments, satisfied the housing requirements, but were quite alien to the area. In addition to these incongruities of dwelling type, the layout of the new development was far removed from the building line and deep strip-plots characteristic of this type of area. Small culs-de-sac, which give access to redeveloped plot tails, are dominated by parked cars as there is little space for off-street parking. Dwellings have been squeezed in to achieve the maximum density, and consequently there are a number of cases in which first-cycle houses, constructed along the original building line, contrast with new dwellings built at right angles to the street frontage within the plot tail, closely adjoining the rear of the original dwelling. This has led to problems of overlooking and reduced daylight, in addition to exaggerating the effect of contrasting styles, scales and layouts (figure 9.6). Furthermore, the form in which a number of the original detached houses have been retained has been far from satisfactory. Their conversion to flats was undertaken purely with the intention of providing a large number of small dwelling units. The contribution that the surviving detached houses would make to the urban landscape was not a consideration (figure 9.7). They were retained owing to financial expediency. Many façade designs and details were destroyed to provide additional entrances, and new plumbing and windows. The layout of car-parking areas was ill considered, landscaping was of poor quality, and the level of building maintenance was low.

A major reason for this thoroughly unsatisfactory outcome in the landscape was the motivation for the redevelopment. It was a political

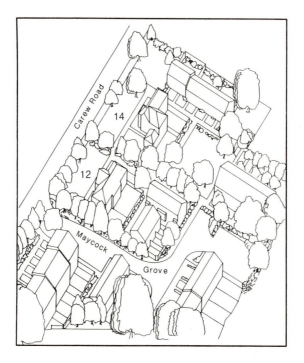

Figure 9.6 Axonometric plan of Maycock Grove, Northwood (redrawn from a planning application).

Figure 9.7 Contrast between retained first-cycle detached house and infill terrace houses, 12 Frithwood Avenue, Northwood (redrawn from a planning application).

response to the desire for social restructuring and the perceived need for more council houses. The impact on the mature, high-quality urban landscape, much of which has subsequently been designated as a Conservation Area, was never a consideration. When the Council came under Conservative control following the local elections in 1977, it promptly withdrew all planning applications for municipal housing developments and sold the sites to private developers.

Chesham Road, Amersham

The municipal housing sites in the Northwood study area are unusual in the extent to which party politics have played a direct role. The Amersham study area illustrates the much more common situation, in low-density residential areas, of an LPA seeking to restrain individual owner-occupiers from maximizing the development value of their property. Here the LPA is, in an important respect, the guardian of existing use values. The most direct conflict is between the LPA on the one hand and developers and owners of land with development potential on the other. The working-out of that conflict is primarily at the scale of individual plots or groups of plots, although somewhat wider locational considerations, notably relating to traffic flows, play a part.

The Amersham study area was part of a much larger area for which the LPA had produced only the broadest of plans. The relevant formal plan was a town map, approved in 1958, on which primarily residential areas were differentiated only in terms of gross population densities calculated for areas much larger than the study area. By the end of the study period, the local plan envisaged in the Town and Country Planning Act of 1971 was still in preparation.

At the end of the 1950s, Chesham Road, the main thoroughfare through the study area, was characterized by large, Edwardian, detached houses in sizeable gardens. On the west side of the road (figures 9.8A and 9.9), which is the subject of attention here, there were practically no other house types. The plots here were to become, over the next thirty years, the subject of no less than fifty planning applications (ten of them leading to appeals) for the construction of new dwellings. These bare statistics provide no more than a suggestion of the interactions between owner-occupiers, developers, LPA and central government that were to result in a new landscape that was primarily a product of conflict and scarcely at all a product of planning.

When, in 1959, the owner-occupier of one of the houses (located just to the south of the area shown in figure 9.8A) applied for outline planning permission to replace his house with twenty-four flats, the LPA was largely unprepared. It cited as a reason for refusing the application the fact that the site was not programmed in the town map for redevelopment – an objection that could have been raised for practically any residential site in the town. The County Highway Authority stated that 'there could be no well-founded highway objection', although it subsequently changed its mind twice about whether one or two main-road accesses were necessary. A second outline application, this time for the construction of sixteen flats, was approved by the Planning Committee two months later, contrary to the recommendation of the planning officers. The LPA was in a state of disarray and lacked a coherent policy.

Further attempts by owner-occupiers and developers to redevelop or infill the west side of Chesham Road at a much higher density were inevitable in view of the increasing pressure on land. Within three years, three outline applications to develop at higher-density individual plots south of Hervines Road, either by demolishing existing houses and redeveloping their sites with maisonettes or by retaining the house and building flats in the garden (figure 9.8B), had been refused. The LPA now argued that the three plots involved, together with two others, should form part of a 'comprehensive redevelopment scheme'. This view was to dominate the LPA's thinking about the future of these five plots

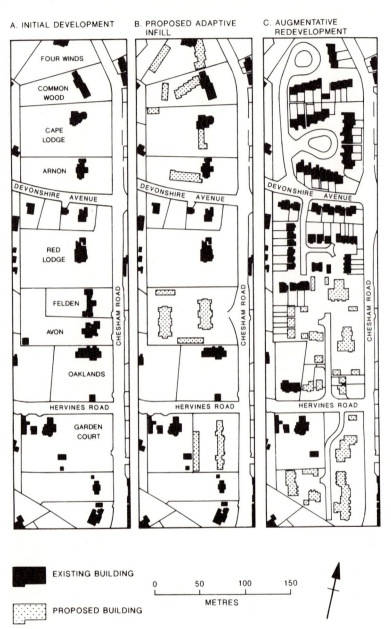

A. INITIAL DEVELOPMENT

B. PROPOSED ADAPTIVE INFILL

C. AUGMENTATIVE REDEVELOPMENT

FOUR WINDS

COMMON WOOD

CAPE LODGE

ARNON

DEVONSHIRE AVENUE

RED LODGE

FELDEN

AVON

OAKLANDS

CHESHAM ROAD

HERVINES ROAD

GARDEN COURT

■ EXISTING BUILDING

PROPOSED BUILDING

0   50   100   150
METRES

244

Figure 9.9 Rear view of Red Lodge, Chesham Road, Amersham, *c.* 1974, based on a photograph in the sale particulars of A.C. Frost & Co., Estate Agents.

and the eleven to their immediate north. Within a month of the submission of the last rejected application, an outline application was submitted by the owners of four of the plots south of Hervines Road for a co-ordinated development, but this was rejected on the ground that no access should be permitted to Chesham Road from the site despite the fact that the County Surveyor raised no objection to such access.

Meanwhile attempts had begun to be made to acquire permission to develop more intensively the garden of Cape Lodge, north of Devonshire Avenue (figure 9.8A, B). Outline permission was sought to add three dwellings alongside the existing house to form a 'courtyard'

Figure 9.8 Main phases in the development and attempted infill and redevelopment of sites in Chesham Road, Amersham. Proposals are omitted in cases in which planning applications were not accompanied by drawings. In most cases, the scheme shown is one of two or more submitted, sometimes over a period of several years. A: Initial development (mainly pre-1930). B: Proposed adaptive infill (mainly 1960s). C: Augmentative redevelopment (mainly 1970s and 1980s) – proposed buildings are those approved or under consideration by the LPA at the end of 1987. (Compiled from planning applications.)

(Whitehand 1990a: 92–5, 1990b: 377–8). Although two not dissimilar proposals had been approved in the vicinity in 1960 (figure 9.10) and 1961, the latter relating to a site on the opposite side of Chesham Road but about 150 metres to the south, this proposal was rejected by the Planning Committee, despite a recommendation of approval by the County Planning Officer, on the grounds that it would cause an increase of traffic on an already busy road and that the site should either be part of a comprehensive scheme incorporating adjoining land and without direct access to the main road, or accommodate only one additional dwelling. There followed in 1965 and 1966 seven outline applications, all from owner-occupiers, relating to the four plots immediately north of Devonshire Avenue (for example, at Four Winds – figure 9.8A, B). Appeals to the Minister of Housing and Local Government against the LPA's refusal of the first two of these applications were dismissed, but the Minister considered it unreasonable to delay development until all the plots of land became available that would be necessary for a comprehensive redevelopment. It became clear to the four owners that co-ordination of their proposals was likely to increase the chances of

Figure 9.10 Front view of a conversion of an existing detached house (to right) to form part of a courtyard development of flats and maisonettes, Chesham Road, Amersham (planning application approved 1960; photograph 1988).

gaining permission, and eventually a number of schemes in which court-yards shared a single access road were approved in quick succession between August 1966 and March 1967. Existing houses and, to a large extent, existing boundaries were to be retained (Whitehand 1990b: 380). Thus, unlike south of Hervines Road, the scene was set for new building to take place essentially within the existing framework of plots and houses ('adaptive infill' in figure 9.8B), consistent with a develop-ment that had recently occurred on the east side of Chesham Road and one on the west side of Chesham Road to the immediate south of the area shown in figure 9.8.

Between Devonshire Avenue and Hervines Road, however, events took a somewhat different course, closer to that already described south of Hervines Road. Concerned by the barrage of applications for permis-sion for piecemeal infill or redevelopment by individual owner-occupiers to the north and south, at a time when comprehensive redevelopment was fashionable, the LPA was resistant to the suggestion of almost any form of piecemeal change. Even an application for an additional pedestrian access to one house (Felden), and another for a temporary site for a caravan adjacent to a house that had already been converted to a nurses' home (Oaklands) were refused. Outline applications for major change followed in 1971, when two owner-occupiers applied for permission to build on their combined sites, first, twenty-two flats and, second, when the application to build flats was refused, twelve detached houses. One of the two alternative schemes included in the application for permis-sion to build flats would have been more appropriate near the centre of a major city (figure 9.11). As was recognized by the LPA, it was quite out of character with the quasi-rural character of Chesham Road and the type of adaptive infill that had, after a long struggle, been approved north of Devonshire Avenue. But the reasons for refusal that had now become a stock response, not only to this application and the applic-ation to construct twelve detached houses, but to any application for permission to undertake development along the west side of this stretch of Chesham Road, were the need for comprehensive redevelopment and the prematurity of development until road improvements had taken place that would relieve traffic congestion on Chesham Road.

In 1973 the owner-occupiers of all four main plots between Devon-shire Avenue and Hervines Road sought to overcome these objections with a scheme for the entire area between these two roads, in which there was no direct access to Chesham Road. This scheme was recom-mended for approval by the planning officers and there was no objec-tion from the County Council Highways Department. Nevertheless, the

Figure 9.11 Front view of a block of flats proposed in 1971 at Avon and Felden, Chesham Road, Amersham (redrawn from a planning application).

Planning Committee refused permission on the ground that it would exacerbate the already unsatisfactory traffic conditions on Chesham Road, a decision that was upheld by the DoE Inspector at a subsequent appeal. Since the only highway changes of which there was a realistic prospect were unlikely to provide major alleviation of traffic congestion on Chesham Road, this last decision meant that more intensive residential development of these plots and those to their immediate south was unlikely to gain approval in the near future.

There followed a series of applications by the owner of one of the plots between Hervines Road and Devonshire Avenue, Red Lodge (figure 9.9), which was used at the time as a nurses' hostel, for a succession of changes of use – first to offices, then to an hotel, and finally to a 'teaching establishment' – and for the construction of flats, in one case aged persons' flatlets. Apart from the proposal to build flats, which was withdrawn, all the applications were refused on the grounds that the land was allocated for residential use on the town map. An outline application for permission to construct a detached house in the garden was subsequently refused on the grounds that it constituted piecemeal development and that it was premature until the future use of the existing house was known. A consequence of this was a further outline application simultaneously to construct a new house and convert the existing one back to residential use. This, an outline application to construct fifty-eight flats (for aged persons) and an outline application to build a detached house on an adjacent plot fronting on to Devonshire Avenue were refused primarily on the grounds that they constituted piecemeal development. Thus, like the plots south of Hervines Road, those between Devonshire Avenue and Hervines Road remained essentially frozen in their existing form, though the houses and gardens were physically deteriorating as continuing uncertainty about their future discouraged investment in maintenance.

A consequence of this series of events was that much of the west side of Chesham Road under investigation suffered from planning blight, albeit that many of its symptoms were screened from public view by the substantial hedgerows fronting the road. Even in the case of the plots north of Devonshire Avenue, for which permissions had been acquired for adaptive infill in the mid-1960s, the landscape outcome remained uncertain for many more years. Actual development ultimately involved the participation of developers, who introduced a perspective quite distinctive from that of individual owner-occupiers, or trustees acting on their behalf, who had been the instigators of previous schemes. Two developers (prospective purchasers) interested in the potentialities of the area, but not in the intricate scheme of interconnected courtyards that had evolved piecemeal, submitted applications for the comprehensive redevelopment of as much of all four sites as would be consistent with the desire of one or two owners to retain parts of their plots (Whitehand 1990b: 377–80). The applications were refused. Ironically, in the light of the LPA's reaction to earlier courtyard proposals, a major reason given was that a courtyard-type development was considered to be most appropriate. However, after further proposals for 'estate' developments by several developers who had in turn purchased, or were prospective purchasers of, part of all of the site, two similar estates of terraced houses in an Anglo-Scandinavian style were built in the late 1960s and early 1970s. Thus an urban landscape was created that was quite out of character, not only with its predecessor and the surviving landscape surrounding it, but with all the schemes put forward by the owner-occupiers.

Meanwhile, the remainder of the west side of Chesham Road continued essentially unchanged, except for the physical deterioration of houses and gardens as owners minimized maintenance in the face of uncertainty. It was not until 1979 that a further attempt was made to obtain planning permission for the plots between Devonshire Avenue and Hervines Road. On this occasion a development company, a prospective purchaser, sought permission to build sixty-three flats and thirty-one 'mews' houses. Permission was refused on the grounds that the density proposed was excessive and, as in the case of previous attempts to develop this land, that development would be premature until traffic problems in Chesham Road had been solved. Again, this second objection was despite the fact that the authority responsible for highway matters, the County Council, raised no objection. It was not until 1984 that the same developer eventually acquired planning permission, this time for the construction of sixty-four dwellings. Two years later the developer obtained permission for slightly more dwellings, and began development of the northern part of the site. The southern part, its original dwellings demolished and their gardens cleared of trees, other than those near the site boundaries, was still awaiting development at the end of the 1980s, as a slump in the housing market brought a cessation in building activity.

The plots south of Hervines Road remained without planning permission even longer, until 1988. Attempts by individual owner-occupiers to obtain permission for developments within individual existing plots on the southern margin of figure 9.8, in 1970, 1983 and 1987, all failed. The Planning Committee refused another co-ordinated application (by the owner-occupiers of three plots) in 1987, despite a recommendation of approval by their planning officers. Eventually the same developer who had started to develop the plots between Devonshire Avenue and Hervines Road contrived a scheme that the Planning Committee felt unable to resist (figures 9.8C and 9.12), faced as they were by the disaffection of their planning officers and an appeal against their refusal of the previous, higher-density proposal. However, the site had scarcely been divested of two of its three houses and most of its shrubs and trees when a slump in the housing market, at the end of the 1980s, left the site with a road and dwelling foundations as the only significant tangible products of 'planning' activity over some thirty years.

Thus, after three decades of attempts to promote and frustrate development, the west side of Chesham Road was essentially as shown in figure 9.8C. Roughly one-half of it was builders' sites on which work had been suspended. The accompanying despoliation of the landscape made the previous state of minimum maintenance of houses and gardens seem a comparatively minor blemish on the landscape. Only one of the substantial houses present at the end of the 1950s had survived. The sites reflected the architectural styles and building types fashionable at the time of redevelopment – Anglo-Scandinavian terraced houses north of Devonshire Avenue and small, pseudo-vernac-

Figure 9.12 Front view of part of a development proposed in 1987 at Garden Court, and two adjacent plots, Chesham Road, Amersham (redrawn from a planning application).

ular, detached houses immediately south of Devonshire Avenue. The state of Chesham Road as a thoroughfare, the putative improvement of which had been the pretext for resisting proposals for development for over a quarter of a century, was, apart from the introduction of mini-roundabouts at two road junctions, essentially unchanged physically at the end of the 1980s (although there had been a great increase in the amount of traffic) from what it had been in the 1950s, when the pressure for more intensive development began.

At no point in the 'planning history' that has been described did the landscape constitute a significant consideration for the LPA. Although more recent reports by case officers occasionally made passing mention of trees and architectural styles, the overriding consideration was the prevention of development of any kind. On matters of density, the LPA was well aware that sites close to the town centre might reasonably be developed at a high density. But local residents, other than owner-occupiers of potential development sites, were vehemently opposed to change of any kind and, if change was irresistible, the prime concern of most of them was to minimize the number of new dwellings

251

constructed. Perversely, in the case of the sites north of Devonshire Avenue, which were the only ones on which development had been completed by the end of the 1980s, the density of dwellings that ultimately emerged was higher than that in any of the schemes of adaptive infill that had been proposed earlier. Furthermore, if similar schemes of infill, in which existing houses were subdivided to form part of a scheme of flats or maisonettes, had been permitted along the entire length of Chesham Road that has been examined, it is arguable that a landscape much more in keeping with that previously existing, retaining the majority of trees and hedgerows, would have resulted. The effects of the small amounts of extra traffic thereby generated need have been no more serious than in the case of the schemes actually completed, or likely to be completed, in the next revival of the housing market.

The history of attempted urban landscape change that has been uncovered in Chesham Road was dominated by the economics and fragmentation of land ownership. Only in the case of the land north of Devonshire Avenue was the eventual co-operation between owner-occupiers in co-ordinating their schemes sufficient on its own to gain planning permission, and even there completely different schemes were eventually constructed after developers acquired the land. Elsewhere, the intervention of a developer proved a significant factor in wresting planning permission from the LPA.

Thus urban landscape change was essentially the outcome of a battle of attrition in which the LPA delayed markedly the onset of a more intensive land use. The landscape itself, though in practice a crucial element in the environment of those living in the area, was never a significant issue in the conflict. One of the few beneficial aspects of the series of events that has been described is that by resisting development the LPA succeeded in deferring a substantial part of it until architectural styles more sensitive to the existing landscape had come into fashion.

Brookdene Drive, Northwood

Watford Road, the main road through the Northwood study area, demonstrates, at a similar scale to the Chesham Road case study, both conflict and co-operation. In 1947, this stretch of the Watford Road contained entirely detached dwellings in large grounds (figure 9.13A). The infilling of rear gardens with additional dwellings became common in this area during the 1950s. It was the intention of the LPA, Ruislip-Northwood Urban District Council, to promote comprehensive development schemes for such areas as a response to the County

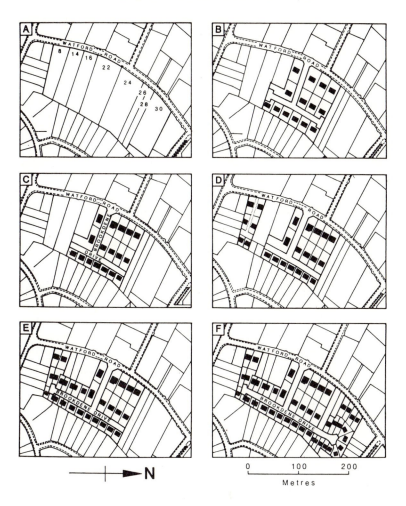

Figure 9.13 Phases of the development of Brookdene Drive, Northwood (redrawn from planning applications).

Development Plan (County of Middlesex 1947), which identified a significant shortfall in housing. As the majority of dwellings were in sound condition, the emphasis of the LPA's policy was on comprehensive infilling rather than redevelopment. The initiative for comprehensive development of the rear gardens was taken by a local developer, Paul Hurst Developments, who on a number of occasions stimulated development in this area by approaching owner-occupiers who had not previously sought to develop their property. The LPA influenced the form of development by liaising with Hurst. Applications by owner-occupiers that might prejudice a comprehensive scheme often met with refusal. Hurst's proposal was for 14 detached houses around an access road (figure 9.13B). Each plot was to be sold to an individual, who would choose a house design from the developer's portfolio and make a detailed planning application. This is a common tactic of developers in such areas, as Jones (1991) has shown. Subsequently, successive applications for permission to construct first 13 houses and, secondly, 40 flats in ten blocks were refused on the grounds that they would result in overdevelopment. Following further negotiations, an application for 17 detached houses was submitted (figure 9.13C), with an undertaking to invest money in landscaping and tree preservation. Some form of agreement was evidently reached between the LPA and developer, in which the developer obtained permission for a higher density of single-family detached houses in return for providing a comprehensive scheme and better environmental quality.

Negotiations followed with adjacent owners, notably those of 8–14 Watford Road. Conflict occurred when the owner of no. 14, reluctant to sell land to the developer, and despite the LPA's views on comprehensive schemes, pursued a private plan to develop the rear of his plot. Although this proposal received permission, access to Brookdene Drive was not agreed and the scheme was not built (the two plots proposed are shown on figure 9.13D). The owners of nos 8 and 10 Watford Road, awaiting a link to the Brookdene scheme, obtained permission to redevelop their plots with six detached houses using a temporary access to Watford Road (see figure 9.13D, on which the access is shown by dotted lines). Although this access was unpopular with the Highway Authority, the LPA was anxious to see the scheme complete. Three years passed with little action, until in 1965 no. 14 was sold to the developer, apparently following the death of the owner. An application for permission to construct four more houses was made and the southern link was completed (figure 9.13E). An application to extend Brookdene Drive to the north and to construct 15 detached dwellings was made in 1965 with the agreement of the owners of nos 26, 28 and 30 Watford Road. A later application to demolish the existing houses and build 17 dwellings was also approved, as was a third proposal to construct 17 dwellings, including the retention of nos 26 and 28. This completed the Brookdene Drive scheme (figure 9.13F).

In all, 47 detached dwellings were constructed on the plots of 8 first-cycle houses, three of the original houses being retained. It is a notable example of a scheme that was comprehensive in intent, but which proceeded in a piecemeal manner; there was apparent co-operation between the developer and the LPA, and landowners were persuaded to join the scheme, although there was some conflict and delay owing to the intransigence of one owner. The result was a development that satisfied the LPA's desire for renewal of the housing stock of the area, achieving a unified plan and high standards of landscaping, whilst limiting the number of vehicular accesses to the busy Watford Road. The informal relationship between LPA and developer was particularly evident, and was presumably aided by Paul Hurst Developments being an established local firm. Schemes by other prospective developers were refused on the grounds that they did not contribute to the comprehensive redevelopment scheme. Yet, in many cases, adaptation of these proposals would have ensured conformity with the LPA's requirements.

Clifton Road, Tettenhall

The final case study is concerned with the micro-scale, the principal focus of attention being the relationship between individual site owners and an LPA in an area of less frenetic change. Again, the protracted pre-development process led to eventual urban landscapes quite different from those originally proposed.

Clifton Road in Tettenhall, some 2 miles from Wolverhampton town centre, runs along the crest of a sandstone ridge, from which there are views over the Smestow valley and, in the distance, Wolverhampton. This hilltop site was laid out in the mid-Victorian period for large villas in extensive well-wooded grounds, and several of these plots have been subject to redevelopment proposals over the last three decades, although for over half of this period they have been part of a Conservation Area. In the majority of cases, the original proposal was for a block of flats; in all but one case, however, the eventual development was of detached houses, giving a far lower density of dwellings.

In the case of the plot at the rear of 28–36 Clifton Road there was a gestation period of nearly seven years, the main conflict being between a potential developer and a neighbouring landowner. Following an initial permission for a 'two-storey block of flats with one flat per storey' in 1971 (figure 9.14A), a local property developer, who had purchased the gardens of nos 28–36, submitted a proposal for two blocks, each of three flats, in 1972 (figure 9.14B). The developer's agent wrote to the LPA stating that

Figure 9.14 Site at the rear of 28–36 Clifton Road, Tettenhall. A: Site of application 295/70, 'two-storey block of flats, one flat per floor'. B: Site of application 2084/72, 'six flats in two blocks'. C: Site of application 618/74, 'flat development'. D: Site of final development, of two detached houses, in 1977. (Redrawn from planning applications.)

My client has made considerable effort to purchase the small piece of land at the front of the site to make a more comprehensive development, but to no avail. It is therefore considered that due to the shape of the site the best form of development would be a few small select flats.... It is considered that to develop this land satisfactorily with houses would result in one plot only, thus sterilizing a large area of developable land.

(Planning files, Wolverhampton Metropolitan Borough Council)

The intransigence of the owner of no. 26 in not selling a small piece of land fronting this site led to a proposal for development on a rather tortuous site (figure 9.14C), permission for which was refused on the grounds that the density would be excessive and adjacent properties would be overlooked. By 1974 the same developer and architect had informally proposed two further schemes; one for seven flats in two blocks (about 14 dwellings per acre, or about 35 dwellings per hectare) and another for 12 flats (about 24 dwellings per acre, or about 59 dwellings per hectare). Informal advice was received from the LPA that both were unacceptable, and a proposal similar in scale to that made in 1972, but on a larger site, was submitted. At this time the LPA was considering extending the Conservation Area, and the case officer noted that, although he felt that the application should be refused, he did not see on what technical grounds a refusal could be substantiated in the absence of Conservation Area policies. In fact, the formal grounds for refusal that were eventually given were that the proposed development would neither preserve nor enhance the established character of the area and, in general, would create an unsatisfactory form of infilling. These were contentious grounds for refusal, as the LPA appear to have acknowledged when they noted in a letter to the applicant's agent that an appeal may be submitted 'unless fruitful discussions for your client ensue'.

This case shows a low level of conflict, between a potential developer and a local landowner, which nevertheless delayed development. When the problematic part of the frontage had eventually been purchased by the developer, the eventual development was very different from that originally proposed, with an application for two detached neo-Georgian houses being approved in early 1977 (figure 9.14D). This difference between initial application and eventual development was, however, not directly caused by the conflict itself, but by the delay needed for the resolution of the conflict. The different LPA policies imposed after the Conservation Area designation and the change in economic conditions between 1971 and 1977 were significant factors.

The recent site history of 46 Clifton Road, very close to the previous site, also began with a proposal for a block of flats, this time in 1960. This application was 'deemed refused' by the LPA on the grounds that too little information was provided. A proposal for 12 flats in 1961 was refused, and there is evidence in the LPA's files for an unsuccessful appeal, although neither the LPA nor the DoE Planning Inspectorate possesses copies of a decision letter. A proposal later that year for 20 flats was refused and, when the architect submitted further plans, an appeal was adjourned for the LPA to consider these as a new proposal. These new plans, suggesting 12 flats, were refused in 1962. Again, there is unconfirmed evidence of an unsuccessful appeal. All of these applications were made by an elderly lady owner, not resident in the area but evidently wishing to capitalize on her property. In 1961 she accused the LPA, then the Tettenhall Urban District Council, of having 'prevented me from developing my property from motives of partiality for local agents and building contractors who are interested in my land but will not pay me a fair price for it' (planning files, now held by Wolverhampton Metropolitan Borough Council). She further claimed that an earlier letter from the LPA, which does not survive, had assured her that 15 or 16 flats could be erected on the site, and that 'an impartial architect has assured me that the site would easily yield 20 flats and is ideally suited for the erection of a block of flats of that density'. The owner's aim was apparently to gain the greatest financial return from the site. The succession of proposals, including six-storey blocks of flats in an area predominantly of substantial two-storey detached houses (figure 9.15), paid no heed to the context of the site. The LPA refused these proposals on the grounds that the building of a six-storey block of flats on this site, situated between two-storey dwellings, would be prejudicial to the amenities of those existing dwellings by reason of appearance, noise and additional traffic to the premises, and would be incongruous, creating a strident feature in the street scene. This clearly suggests concern for the character of the urban landscape on the part of the LPA.

By 1971 the elderly owner had died, and her executors submitted a proposal for two blocks of two flats each, which was approved. By 1975 serious thought was being given to the implementation of this permission, but by this time the site had become part of a Conservation Area. This area also had the status of an Area of Special Development Control, with development being restricted to three houses, or four flats, per acre (about seven houses, or ten flats, per hectare). LPA officers were concerned about the 1972 permission for four flats, describing detailed plans submitted in 1975 as 'gross overdevelopment'.

Figure 9.15 Block of flats proposed in 1972 on the site of 46 Clifton Road, Tettenhall (redrawn from a planning application).

Following meetings between the architects and senior planning officers, the architects wrote that 'we have been instructed by our clients to suspend negotiations ... as we are given to understand that they are now in the process of selling the land' (planning files, Wolverhampton Metropolitan Borough Council).

By December 1975 the site had been acquired by the same developer who was involved in the site adjoining 26 Clifton Road. Employing the same architect, he proposed two detached dwellings, similar in appearance to his other development (figure 9.16). Although the style was clearly neo-Georgian in inspiration, he noted that 'we would emphasize that our design ideas should not be related to Neo Georgian. We are against ... the excessive and gimmicky use of glass fibre pillars and suchlike [*sic*]. Our design is simple but elegant and with the high quality of building and landscaping in which our company has a proven record, this would be an outstandingly pleasing development'. The scheme received planning permission and was built, and it is a form of infill development that is more compatible with the existing urban landscape than were the earlier proposals for flats.

The first part of the site history of 46 Clifton Road shows clear evidence of conflict between site owner and LPA, whose aims were evidently incompatible. Later, the designation of the Conservation Area brought about a change of policy on the part of the LPA or, more correctly, since this is not a specifically published policy, on the part of

Figure 9.16 One of a pair of neo-Georgian houses constructed in the late 1970s on the plot of 46 Clifton Road, Tettenhall (photograph by P. J. Larkham, 1985).

its planning officers. The second part of the site history shows considerable accord between LPA and developer, although the LPA case officer did informally note his reservations. He was concerned that a rash of houses of the same design might be built on infill sites throughout the Conservation Area, but there were no design details which the LPA could claim did not conform to their conservation policy. In these circumstances he could only recommend approval for such proposals. But the good relationship between LPA and developer is underlined by the architect's letter to the LPA, which registered appreciation of the way in which the LPA officers 'handled this difficult application in a positive and expedient manner'.

CONCLUSIONS

The cases that have been described are illustrative, and comparable studies of considerably more sites are required before confidence can be placed in their representativeness. Nevertheless, in the context of the wider-ranging analysis of the six study areas with which this chapter began and taking account of the findings of the existing body of related work referred to earlier, a number of conclusions seem to be justified.

First, the timing of attempts to change the landscape is important. In these residential study areas, that timing is heavily influenced by the life cycles and other circumstances of owner occupiers (Whitehand 1990a: 90), but there is a variable, sometimes lengthy, time lapse between attempts to bring about change and actual change. This is influenced to a considerable extent by LPAs. It may mean that developments take place in conditions markedly different from those in which they were conceived (Whitehand and Larkham 1991b). This was certainly the case at Chesham Road between Devonshire Avenue and Hervines Road, and was also true of the sites on Clifton Road.

Secondly, economic factors would seem to underlie some major differences between South-East England and the English Midlands (Whitehand and Larkham 1991a, b). The results of the comparison of these two regions are consistent with processes heavily influenced by pressure on land. Development control officers tend to argue that the economics of proposals is not their business. But it is clear from all six study areas examined in this chapter that economic considerations are so central that it makes little sense to respond to proposals as if the financial returns to the developer were immaterial.

Thirdly, many more urban landscapes exist on paper than ever come into being on the ground. The successions of proposals rarely consist of progressions towards improved solutions. A great deal of energy, some of it creative, is clearly wasted in this process. The predevelopment process, from initial discussion and application onwards, is inordinately time consuming for both applicants and LPAs (Whitehand and Larkham 1991b). Whether the quality of the outcome would be worse if development or non-development followed a single planning application processed by the LPA within the statutory eight-week term is a matter for speculation.

Fourthly, in these originally low-density residential areas there is little that could be described as 'management' of the urban landscape, in the sense in which that activity is envisaged by Conzen (1966, 1975, see also chapter 1 of this volume). Nor, indeed, are there many developments that could be described as 'planned' outcomes (Whitehand 1990b). The effects of local-plan documents have been quite small at the scale of individual sites, even in designated Conservation Areas (Whitehand and Larkham 1991b).

Fifthly, if LPAs are judged by their actions in these study areas, then for them highway matters and density are far higher priorities than the visual environment. The forms of developments, particularly their architectural styles, have tended to be afterthoughts, when battles over density and access have been concluded (Whitehand 1990b); an order of priorities generally supported at appeals. Attempts by applicants to meet or obviate density and highway objections have often been at a cost to the urban landscape in terms of both the relationship between buildings and the survival of the trees and hedgerows that are responsible for much of the character of low-density residential areas (Whitehand 1989b: 419). As far as the urban landscape is concerned, there is a price to pay in lost cultural assets.

Sixthly, inadvertent outcomes are widespread. Actual changes to the urban landscape are commonly substantially different from those originally envisaged by the parties involved in the development process (Whitehand 1990b). The redevelopment of sites in Clifton Road as detached houses rather than as blocks of flats is an example.

Seventhly, the dominant impression of the decision-making process leading to development or non-development is that it consists of a number of poorly co-ordinated activities in which expediency plays a major part (Whitehand 1990b: 389). Even successive applications by the same developer for the same site may be unrelated in terms of the physical form of proposed developments. Later schemes often reflect attempts to achieve higher densities, as seems to have been the case with some applications by Paul Hurst Developments to undertake developments in Watford Road, Northwood. LPA case officers tend to be collectors and transmitters of views and instructions rather than integrators. They should have major roles in ensuring that awareness of the relationship of the character of proposed developments to the existing urban landscape is not subordinated to a concern with technicalities. LPA case officers should be central to the activities of urban-landscape management described by Conzen (1966, 1975).

Eighthly, infill and piecemeal redevelopment often requires the employment of particular ingenuity if the visual environment of neighbouring occupiers and cumulatively the visual environment of the larger community are not to be adversely affected. The need for that ingenuity is likely to become greater as the shortage of land leads to the redevelopment of medium-sized gardens becoming profitable (Whitehand and Larkham 1990a: 64). The awareness of the importance of the visual environments of the study areas grew throughout the period under investigation, with in some cases the designation of Conservation Areas being associated with changes in policy and changes in LPA attitudes towards development proposals. The dismay of the LPA at the thought that the existing planning permission for the construction of flats at 46 Clifton Road might be implemented, to the detriment of the newly designated Conservation Area, is indicative of this change.

Ninthly, more frenetic activity in South-East England, in comparison with the English Midlands, does not necessarily entail more obtrusive development. Small detached houses, such as are characteristic of recent suburban infill in the Midlands, may be more obtrusive in the urban landscape than flats. For example, the effects of the insertion of numerous separate private driveways to individual dwellings can devastate an existing hedgerow that it might have been possible to preserve if a single block of flats had been constructed. From this standpoint the vociferous opposition of local residents to flats is misplaced (Whitehand 1989a; Whitehand and Larkham 1991a: 63). It is to be hoped that recent successful attempts to design blocks of flats that are similar externally to large, detached houses will lessen this opposition.

The major theme of the entire process of redevelopment and infill, as revealed in this study, is that of conflict. Conflict arises from the very nature of the development process as institutionalized in UK law. The agents active in the process tend to have different viewpoints, ranging from developers seeking to change the landscape from motives of profit, to residents wishing to minimize change, much of which is inimical to the maintenance of the existing use value of their property. Of course, some residents may become developers: positions within the system are not static over time and, far from devaluing a property, as is often claimed, a new development may prompt a neighbouring owner to realize the development potential of his own site. Conflicts are institutionalized in the quasi-legal process of working out differences of opinion – and, sometimes, of fact – in the operation of the development-control system at the local level and on appeal to the Secretary of State for the Environment. Throughout this process, the roles of local authority case officers emerge as being potentially important. The central position of case officers within procedures for processing applications makes them better equipped than other parties to develop an integrated view. Yet they are in awkward positions in practice, being intermediaries between developers, objectors and planning committees. There have been accusations from developers, residents and other agents, of wilful neglect, even misconduct, on the part of case officers and planning committees, although these are rarely pursued or substantiated. The recent examination of 122 contentious cases in Brent (Bar-Hillel 1991) is unusual, and found insufficient evidence to justify conclusions of corruption or malpractice. Similar accusations would be hard to substantiate from the many hundreds of applications examined in the course of this study. What is clear is that case officers have a role in practice that is insufficiently integrative. Reactions to proposals in terms of the mechanical application of rules about densities and highways are anathema to the urban landscape. Case officers should have a major role in ensuring that the total character of developments and their relationships to existing urban landscapes are central to the development process.

ACKNOWLEDGEMENTS

The research described in this chapter was funded by the Leverhulme Trust and the Economic and Social Research Council. The authors are indebted to Paul Booth for help with the preparation of figure 9.1. Some of the case studies employed in this chapter have been discussed, in different contexts, in other publications.

REFERENCES

Amos, C. (1991) 'Flexibility and variety – the key to new settlement policy', *Town and Country Planning* 60, 2: 52–3.

Bar-Hillel, M. (1991) 'Independent report questions Brent planning discrepancies', *Chartered Surveyor Weekly* 15/22 August: 14–15.

Bibby, P. and Shepherd, J. (1991) *Rates of Urbanization in England 1981–2001* London: HMSO.

Booth, P. N. (1989) 'Owners, solicitors and residential development: the case of a Manchester suburb', unpublished M.Phil. thesis, Department of Geography, University of Birmingham.

Bromsgrove District Council (1984) *Guidelines for the Conversion of Agricultural Buildings*, Bromsgrove: Bromsgrove District Council.

Chartered Surveyor Weekly (1988) 'More homes for Surrey: Ridley', *Chartered Surveyor Weekly*, 7 April: 25.

Conzen, M. R. G. (1966) 'Historical townscapes in Britain: a problem in applied geography', in House, J. W. (ed.) *Northern Geographical Essays in Honour of G. H. J. Daysh*, Newcastle upon Tyne: Oriel Press.

Conzen, M. R. G. (1975) 'Geography and townscape conservation', in Uhlig, H. and Lienau, C. (eds) 'Anglo-German Symposium in Applied Geography, Giessen-Würzburg-München, 1973', *Giessener Geographische Schriften* 1975: 95–102.

Cornish, D. (1988) 'Second-cycle change in the residential townscape: Sale, Manchester, 1947–1987', unpublished BA dissertation, Department of Geography, University of Birmingham.

Damesick, P. J. (1986) 'The M25 – a new geography of development? I. The issues', *Geographical Journal* 152, 2: 155–60.

Department of the Environment (1987) *Development Control Statistics: England 1983/84–1985/86*, London: DoE Land and General Statistics Division.

Goodchild, R. and Munton, R. (1985) *Development and the Landowner: an Analysis of the British Experience*, London: Allen and Unwin.

Hamnett, C. (1986) 'The changing socio-economic structure of London and the South East, 1961–1981', *Regional Studies* 20: 391–406.

Holland Report (1965) *Report of the Committee on Housing in Greater London* Cmnd 2605, London: HMSO.

Jones, A. N. (1991) 'The management of residential townscapes', unpublished PhD thesis, School of Geography, University of Birmingham.

Larkham, P. J. (1990) 'The use and measurement of development pressure', *Town Planning Review* 61, 2: 171–84.

London Borough of Hillingdon (1973) *Frithwood Avenue/Carew Road/Old Northwood Policy for Controlling Development*, Action Area Plan, Uxbridge: London Borough of Hillingdon.

Pompa, N. D. (1988) 'The nature and agents of townscape change: South Birmingham 1970–85', unpublished PhD thesis, Department of Geography, University of Birmingham.

Punter, J. V. (1989) Development control – case studies, in Hebbert, M. (ed.) *Development Control Data: a Research Guide*, Swindon: Economic and Social Research Council.

Simmons, M. (1986) 'The M25 – a new geography of development? III. The emerging planning response', *Geographical Journal* 152, 2: 166–71.

Tym, R. and Partners (1987) *Land Use for Residential Development in the South East: Summary Report*, London: Roger Tym and Partners.

Whitehand, J. W. R. (1967) 'The settlement morphology of London's cocktail belt', *Tijdschrift voor Economische und Sociale Geographie* 58, 1: 20–7.

Whitehand, J. W. R. (1988) 'The changing urban landscape: the case of London's high-class residential fringe', *Geographical Journal* 154, 3: 351–66.

Whitehand, J. W. R. (1989a) *Residential Development under Restraint: a Case Study in London's Rural-Urban Fringe*, Occasional Publication 28, Birmingham: School of Geography, University of Birmingham.

Whitehand, J. W. R. (1989b) 'Development pressure, development control and suburban townscape change', *Town Planning Review* 60, 4: 403–20.

Whitehand, J. W. R. (1990a) 'Makers of the residential townscape: conflict and change in outer London', *Transactions of the Institute of British Geographers* NS 15, 1: 87–101.

Whitehand, J. W. R. (1990b) 'Townscape management: ideal and reality', in Slater, T. R. (ed.) *The Built Form of Western Cities*, Leicester: Leicester University Press.

Whitehand, J. W. R. and Larkham, P. J. (1991a) 'Housebuilding in the back garden: reshaping suburban townscapes in the Midlands and South-East England', *Area* 23, 1: 57–65.

Whitehand, J. W. R. and Larkham, P. J. (1991b) 'Suburban cramming and development control', *Journal of Property Research* 8, 2: 147–59.

Worcestershire County Council (1975) *Worcestershire Structure Plan*, Worcester: Worcestershire County Council.

10

RECENT CHANGE IN TWO HISTORICAL CITY CENTRES: AN ANGLO-SPANISH COMPARISON

J. Vilagrasa

INTRODUCTION

This chapter is concerned with the results of empirical research on post-war change in city centres in England and Spain. Methods developed for the study of morphological change in English towns (Whitehand and Whitehand 1983, 1984) have been applied in Spain, where there exists a source of detailed data similar to that held by local authorities in the UK. This is the first research applying these methods in an international comparison. Rates and types of building replacement, and alterations to buildings of recognized architectural or historical merit (hereafter 'historical buildings'), in the commercial cores of Worcester (UK) and Lleida (Spain) are examined, from the end of the Second World War in Worcester and from the Spanish Civil War in Lleida.

The study areas

The cities were chosen because of their essentially similar characteristics of form and function. In size, both have populations of about 100,000; in function, both are important regional centres and cathedral cities, with considerable tourism potential; and in form, both are centred on one bank of a major river, with relatively little development on the other side.

Lleida (Lérida) is located on the northern bank of the River Segre. It was a prosperous town in the Roman period, and became an episcopal see in the Visigothic period. Occupied and rebuilt by the Moors in the eighth century, it was reconquered in 1149, and repeatedly besieged and occupied by the French in the eighteenth and nineteenth centuries

(Gutkind 1967: 307). The city is dominated by the hilltop citadel and cathedral (la Seu Antigua), which was consecrated in 1278 but was used for military purposes from the eighteenth century until recently. The old city is crowded into the small plain between the steep hillside and the river, but the walled medieval city included a considerable expansion to the west of the citadel, where a *villa nova* was developed at the beginning of the thirteenth century. Several phases of planning are evident, not least in the construction of the main square, the Plaça de Sant Joan (Lladonosa i Pujol 1983: 12–16). These are apparent in the substantial remains of the early church of Saint Joan de la Plaça discovered underneath the present *plaça* in 1976 (Gallart *et al.* 1985: 13–24), and in the successive wall lines following urban expansion, conquest and contraction (Gallart *et al*, 1985: figure 42). The city underwent a building boom in the late nineteenth and early twentieth centuries, particularly for banking and commercial uses along the river front and Carrer de Blondel, and in the construction of wide boulevards (*ramblas*) along the lines of the demolished medieval wall (Lladonosa i Pujol 1983).

Worcester has many similarities to Lleida, although it does not have the physical constraint of hill and narrow floodplain. The city is on the south bank of the River Severn, and is the only crossing-point of the Severn between Gloucester and Bridgnorth. It is thus a key strategic and trading point. There was at least a small Roman settlement here, an episcopal see was established *c.* 680, and the settlement was fortified some two centuries later. There is some debate whether the medieval wall followed the line of this Anglo-Saxon fortification. In its regularity of plots, the medieval town certainly shows evidence of planning, in several phases (see Chapter 3 of this volume). Expansion in the post-medieval period was slow, and even in the mid-eighteenth century there were few more buildings outside the walls than in the first years of the sixteenth century, as a comparison of Speed's map of 1610 and Doharty's of 1741 shows. But the industrial expansion of the nineteenth century, with its associated housing, occurred both within the medieval city and immediately adjoining its walls. 'Factories and housing were crowded together in the centre of the town, between the High Street and the river, along the canal and in the Blockhouse area... [a] depressing jumble of grim, industrial building and bleak rows of houses' (Minoprio and Spencely 1946: 11). Both cities faced considerable pressure in the latter half of the twentieth century to expand their commercial functions, cater for increased tourism, and to become conservation conscious. The results of these pressures are described in this chapter.

Data sources

Both Spain and Britain possess detailed data sources on changes to the urban built fabric, both resulting from the legal requirement to control development, and being sufficiently similar in nature to permit comparative study (Vilagrasa, forthcoming). The Spanish *licencias de obras* or *permisos de obras* contain details of the developer (*promotor*), architect, date and location of the development, and a number of measurements required by the Spanish regulatory system, including frontage, areas and number of floors (Vilagrasa 1984: table 1). Architectural drawings are usually available. This data source has been used in a number of studies of twentieth-century urban change (for example, Vilagrasa 1984; Gomez Mendoza 1986; Alió 1987; Grup d'Estudis Urbans 1987; Larrosa i Padró 1987). The British systems of building and planning applications provide similar data on the agents and types of change, with the planning application files also containing further information in the form of internal memoranda and correspondence leading to the decision of whether to permit or refuse the application. Both British data sources have been used for detailed studies of the form and development of town centres and residential areas during the last decade (Whitehand and Whitehand 1983, 1984; Larkham 1988a). The similarities between the Spanish and British sources have been explored (Vilagrasa forthcoming), and it is clear that they permit meaningful comparisons to be made between the selected city centres.

The analysis

Five significant aspects are examined for each commercial core. First, physical change, relating to new buildings and changes to historical buildings, is analysed. Secondly, the agents involved in the processes of change, particularly developers and architects, are examined. Thirdly, the evolution of urban planning traditions for city centres is considered for both countries, and in particular the implications of planning policies for commercial development and for buildings and areas considered to be historically and architecturally significant are discussed. Fourthly, the succession of architectural styles employed in new buildings is examined, together with the cultural and architectural ideas underlying changes to historical buildings. Lastly, consideration of the previous four points together casts light on the innovation diffusion of the architectural forms that make up the commercial urban landscape. This process reflects the interaction of the various agents of change and the different planning policies in operation.

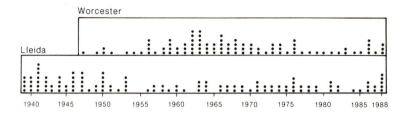

Figure 10.1 Numbers of buildings constructed, and alterations made to historical buildings, in the commercial cores of Worcester (1947–88) and Lleida (1939–88).

THE DYNAMICS OF BUILDING REPLACEMENT

Figure 10.1 shows the numbers of new buildings constructed, and alterations to historical buildings, in the commercial cores of both Worcester and Lleida between the ends of the respective wars and 1988. In Worcester there were 92 developments in 42 years, and in Lleida there were 113 in 50 years. In both cases the mean is approximately 2.2 developments per year.

Despite this apparent similarity, the patterns of change over time were very different. In Worcester there was a distinct slump in the immediate post-war period, a building boom during the 1960s, and a recession from the mid-1970s onwards. These fluctuations correlate with fluctuations in building in general in the UK. In Lleida, a building boom occurred immediately following the end of the war and continued until the early 1950s. There was a lower rate of activity thereafter. There was an inverse correlation here between the city-centre building cycle and the national building cycle.

The analysis of the land uses of new developments in part explains these differences (table 10.1). In Worcester, 90 per cent of new development was for commercial use, with residential use being rare until the 1980s. In Lleida, in contrast, two-thirds of development was for residential use, usually with commercial use at ground-floor level; buildings solely for commercial use became numerically important only during the 1980s. This relates to the Spanish – and indeed southern European – urban tradition of retaining a significant proportion of residential land use in city centres. Spanish housing policies also gave substantial support to residential development between the late 1940s and late 1970s. This makes the inverse relationship between the national

269

Table 10.1 Uses of new buildings in Worcester (W), 1947–88 and Lleida (L), 1939–88.

| | Buildings | | | | | | | | | | | Total buildings | | Total dwellings | |
| | Commercial[1] | | Office[2] | | Residential[3] | | Institutions | | Others[4] | | | | | | |
	W	L	W	L	W	L	W	L	W	L	W	L	W	L
1939–45	–	4	–	2	–	18	–	3	–	0	–	27	–	83
1946–50	2	1	1	1	0	15	1	1	0	0	4	18	0	130
1951–55	1	2	2	1	0	2	1	1	0	1	4	7	0	9
1956–60	8	1	2	0	0	6	2	1	1	0	13	8	0	50
1961–65	9	0	10	1	0	6	2	0	1	0	22	7	0	269
1966–70	12	0	3	1	1	5	1	1	1	1	18	8	16	122
1971–75	4	0	5	3	1	6	0	0	0	2	10	11	2	64
1976–80	2	0	2	3	2	6	1	1	1	0	8	10	46	94
1981–88	7	7	0	5	3	4	2	0	1	1	13	17	23	32

Notes: 1 Buildings entirely in commercial use, including retail shops and those non-retail uses such as building societies and betting shops that possess a retail frontage
2 Buildings entirely in office use together with buildings having commercial uses (see note 1) on lower floors and office uses on upper floors
3 Buildings in commercial use on lower floors but having residential uses on upper floors
4 Workshops and light industrial uses, car parks and other mixed uses

building cycle and the city-centre building cycle in Lleida comprehensible, as the national boom of the 1960s to mid-1970s was essentially a housing boom and involved little development for non-residential uses. This boom mainly affected peripheral areas rather than city centres (Vilagrasa 1987). However, the UK building boom of the 1980s, though associated in particular with the housebuilding industry, was also a commercial boom, with consequent impacts upon city centres.

THE AGENTS OF URBAN CHANGE

The importance of identifying the agents involved in the processes of urban change, and examining their type and provenance, is now well established in British urban morphology (Whitehand and Whitehand 1984; Whitehand 1984; Freeman, 1986; Larkham 1986). The present study draws a distinction between those developers who produce a use value and those who produce an exchange value. Within these categories are secondary distinctions determined by the frequency with which types of agent become involved in the development process. It is significant that one type of 'exchange-value agent', occasional speculative developers, is very important in Lleida but not in Worcester. Figure 10.2 shows the differing incidence of new building and alterations to historical buildings by each type of developer in the two city centres. Four significant points emerge from this comparison. First, of all types of developer, retailers are responsible for the largest number of new buildings and historical building alterations in Worcester (1 in 3), but are relatively unimportant in this regard in Lleida (1 in 10). Secondly, regional or national chains are more important, in terms of new buildings constructed and historical building alterations, in Worcester than in Lleida. In Worcester, the chains are mainly retailers; in Lleida, they are mainly banks, which have built residential blocks in which their offices occupy the ground and first floors. It could be argued that the chains, mainly banks, sometimes produced both use values and exchange values within the same developments in Lleida, producing offices for their own use and flats for sale or lease. Thirdly, new buildings and alterations to historical buildings for institutions are represented in similar numbers in both city centres, but the incidence of new buildings and historical-building alterations for this type of agent is greater in Lleida during the post-war period. Fourthly, in Worcester, speculative developers are well represented from the building boom of the 1960s onwards. In Lleida the trend is markedly different. During the 1940s and 1950s, occasional speculative developers – mainly private landowners as distinct from

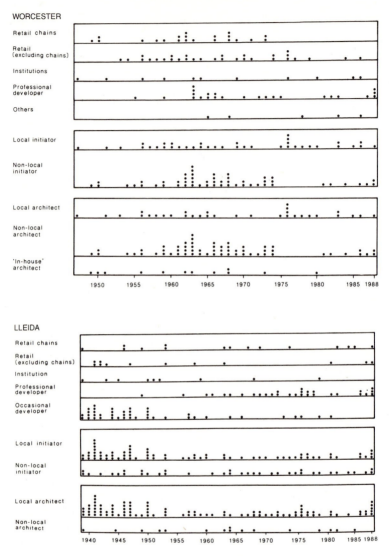

Figure 10.2 Characteristics of direct agents of change responsible for the construction of new buildings and the alteration of historical buildings in commercial cores. A: Worcester (1947–88). B: Lleida (1939–88).

Table 10.2 Provenance of developers and architects engaged in different types of development in Worcester and Lleida.

WORCESTER	Local developer	Non-local developer	London developer	Total developers	Local architect	Non-local architect	London architect	Total architects
Retail chain	0	21	6	21	0	21	6	21
Institution	10	1	1	11	7	4	3	11
Commercial¹	22	8	2	30	18	12	6	30
Professional developer	4	21	19	25	4	21	12	25
Other	3	2	1	5	3	2	0	5
Total	39	53	29	92	32	60	27	92

LLEIDA	Local developer	Non-local developer	Barcelona developer	Madrid developer	Total developers	Local architect	Non-local architect	Barcelona architect	Madrid architect	Total architects
Retail chain	1	14	9	3	15	7	8	6	2	15
Institution	3	6	0	6	9	6	3	0	3	9
Commercial¹	12	0	0	0	12	12	0	0	0	12
Professional developer	17	13	10	0	30	26	4	3	1	30
Investor	38	5	2	0	43	43	0	0	0	43
Other	0	0	0	0	4	0	0	0	0	4
Total	71	38	21	9	109	94	15	9	6	109

Note: 1 Commercial uses here include retail shops that are not part of a chain and those non-retail uses such as building societies and betting shops that possess a retail frontage businesses – were well represented. From the 1960s onwards, these agents have been replaced by professional developers and companies.

The provenance of developers and architects has elsewhere been found to be of significance (Whitehand and Whitehand 1984). In Worcester, two-thirds of new building construction and alteration to historical buildings was by non-local developers (table 10.2). This relates to the importance of the national chains and, more importantly, to the rise of speculative developers, mainly London-based. In Lleida, in contrast, there is a major contribution by local speculative developers. This is mainly due to the numbers of landowners who acted as developers during the immediate post-war years, but there is also a large number of locally based professional speculative developers.

These differences between the characteristics of developers in Worcester and Lleida are associated with different degrees of concentration of the financial, commercial and building activity in the two countries during the study period. In England, a high concentration of activity reflects the importance of national commercial chains and national development companies, even in medium-sized towns. But in Spain, apart from a strong presence of non-local banks, local agents, not all of them professionally qualified, predominate.

The same type of analysis may be carried out for the architects identified in the development process. Non-local developers active in Worcester (principally the retail chains and professional developers) rarely use Worcester-based architects, while in Lleida there is a greater tendency for all developers to use local architects. These differences are explained by the different national traditions of architect/developer relationships. In England, each commercial chain frequently has its own 'in-house' architectural office, working solely for the chain. In Spain, in contrast, it is usual for a development company to employ different independent architectural practices for each development.

URBAN PLANNING AND MANAGEMENT

An understanding of the evolution of planning policy, both national and local, is vital to a complete understanding of morphological change in the two city centres. Figure 10.3 summarizes these changing policies. A comparative framework can be used to understand major plot and street changes, and also to analyse conservation policies. Another important consideration is the different approaches to land-use regulation and building regulation in the two countries.

Land use in Worcester city centre during the study period was mainly

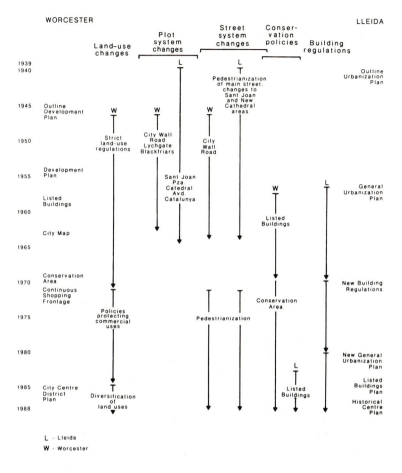

Figure 10.3 Planning policies and documents, Worcester and Lleida, 1939–88.

restricted to commercial uses in the urban core, and institutional and office uses on the periphery of the centre. In practice from the late 1970s, and in planning policy from 1985 (the year of the City Centre District Plan), a more lax land-use regulation was introduced, allowing new residential developments and facilitating the conversion of existing buildings to include housing or offices above the commercial ground floor. In fact, Worcester is one of the successes of the 'Living above the Shop' project encouraged by the Department of the Environment during the 1980s. A result of this less restrictive policy is that the city centre has

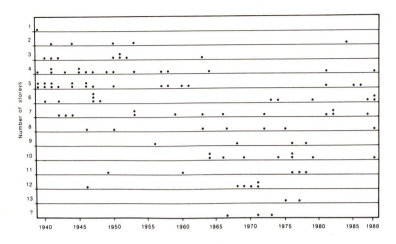

Figure 10.4 Number of floors in new buildings constructed in the commercial core of Lleida, 1939–88.

become more multi-functional. In Lleida, in contrast, commercial, institutional and residential uses have all been permitted throughout the post-war period without specific zoning.

In Lleida, the building regulations in force from 1957 onwards entailed strict control over the number of floors, and thus height, of new buildings, rather than over their use. Figure 10.4 shows a first period, until 1969, when new buildings of four or five storeys were normal, although political interventions allowed a small number of taller buildings. The new building regulations of 1969 allowed more floors, but the height of new buildings was reduced again following the 1979 General Plan. These changing building regulations were a significant element in the transformation of the traditional urban landscape and, most importantly, the skyline, of the commercial core.

Both commercial cores underwent radical transformations in their plot systems and street networks in the post-war period. In Worcester (figure 10.5) these were related to the use of land for commercial and institutional purposes, and to the creation of new roads, principally the City Wall inner ring road. In Lleida, too, changes were associated with commercial and institutional uses (figure 10.6). There was also a major remodelling of the city's main square and the construction of new roads. From the early 1970s in both commercial cores, new planning policies involving pedestrianization were developed and their implementation was begun.

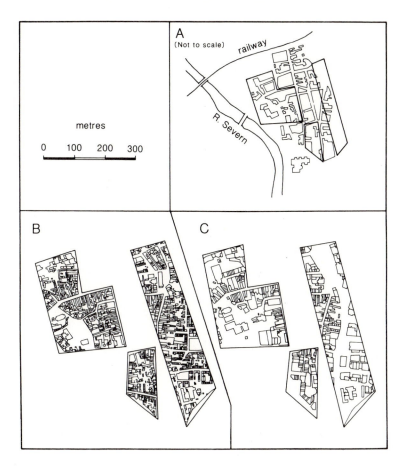

Figure 10.5 Areas of intensive changes to plot and street patterns in the commercial core of Worcester. A: Sketch of location of areas of intensive change. B: Beginning of study period. C: End of study period.

Conservation policies have, however, been radically different in the two centres. In Worcester, a Listed Buildings Plan came into force in the late 1950s and, following the provisions of the 1967 Civic Amenities Act, the city centre was designated as a Conservation Area in 1969. During the 1960s, development control played an important role in conservation, as developers were put under pressure to use neo-Georgian architectural styles in the town centre or, at the least, to change some aspects of the façade designs and materials that they had

277

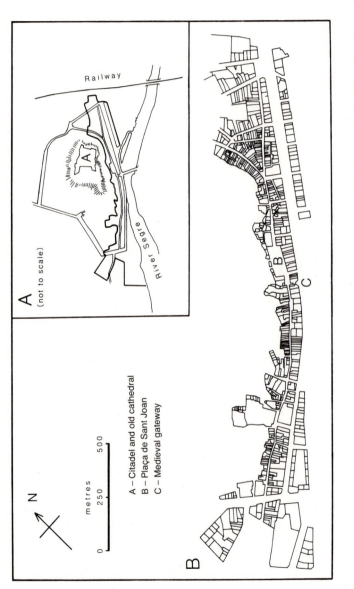

Figure 10.6 Area of intensive changes to plot and street patterns in the commercial core of Lleida. A: Sketch of location of area of intensive change. B: End of study period (note: no detailed plans are available for the beginning of the study period).

proposed. Brick was the preferred material, and the modern-styled long windows, emphasizing the horizontality of buildings, were sometimes rejected. In contrast, the first conservation planning action in Lleida was contained in the General Plan of 1979, rapidly enforced by a Listed Building Plan in 1983 and an Historical Centre Plan in 1985. Development control, as practised in Britain, did not exist during the study period, and no historical or aesthetic criteria for judging new buildings were developed by the Council or the Planning Office. Planning does involve some aesthetic assessment, but a formal development control mechanism for its enforcement is lacking.

ARCHITECTURE AND THE URBAN LANDSCAPE

Examination of the building plans and planning applications in the two centres allows consideration of the architectural styles employed in new buildings during the study period, and their impact on the urban landscape. The changes in styles during the study period are significant. The identification of styles is based on recent architectural criticism (for example Frampton 1980 on the modern style, and Jencks and Chaitkin 1982 on post-modernism) and parallel studies of British commercial cores (including Whitehand 1984, and Larkham and Freeman 1988b). Only three styles were popular in Worcester: neo-Georgian, modern/international and, most recently, post-modern, particularly the neo-vernacular variant. Five styles are evident in Lleida: vernacular, related to the persistence of popular rural building traditions; historicism, mainly of neo-classical form; modern commercial styles, mainly rationalism and expressionism; the modern/international style; and again, most recently, post-modernism.

The changing importance of these different styles is shown in figure 10.7. In Worcester, there was an initial period during which both modern and neo-Georgian styles were employed. After 1969 both styles declined in popularity, being replaced by post-modern styles; this was also the period of a substantial increase in alterations to historic buildings. This change can be explained, in a large part at least, by the effects of conservation planning and development control.

A good example of modernism in Worcester is the City Arcade of the 1950s. The original proposal (figure 10.8) was little altered in the planning process, the main change being the removal of the clock from the first to the third floor, but the tiled façade and window proportions remained unchanged. During the 1960s, however, the neo-Georgian style was suggested by the City Planning Department as an acceptable

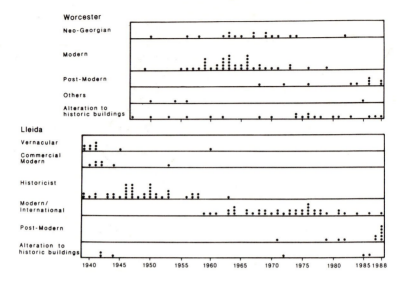

Figure 10.7 Architectural styles of new buildings, and incidence of changes to historical buildings in the commercial cores of Worcester (1947–88) and Lleida (1939–88).

alternative to many modern designs. Figure 10.9 shows this in the changing design suggestions for 29 High Street – from a modern style, copying its northern neighbour, the recently redeveloped City Arcade, to a neo-Georgian style complementing its southern neighbour. Modern buildings were subject to control in terms of both building materials and other aspects of form. Figure 10.10 shows the continued use of brick as a building – or at least cladding – material in Worcester, as opposed to the new materials, particularly glass and concrete, that were becoming prominent in other British town centres at the same time (Whitehand 1984; Freeman 1986). Traditional window shapes remained in common use – a characteristic encouraged by the Planning Department – more so than the modern-styled long window shapes, or glass façades. After 1969, when the bulk of the city centre was designated a Conservation Area, the vernacular aspect of post-modern styling (cf. Jencks and Chaitkin 1982) became dominant in new buildings, which showed similarities to local, pre-seventeenth-century, building forms (albeit that they lacked the distinctive timber framing that had been characteristic of the local region in earlier times) (figure 10.11).

In Lleida, there was a clear dominance of architectural historicism in

280

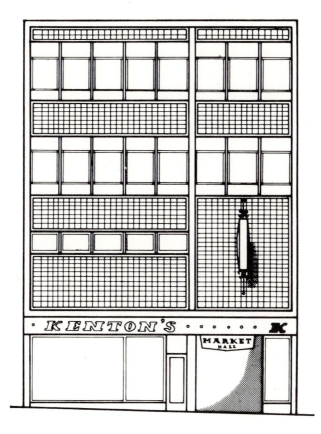

Figure 10.8 Early proposal for City Arcade, High Street, Worcester (redrawn from a planning application submitted in 1955).

new buildings constructed between the end of the Civil War and the late 1950s (figure 10.7), but this was accompanied by examples of vernacular architecture and modern commercial styles. Modern styles predominated in new buildings from the 1960s until the rise of post-modern styles during the 1980s. The dominance of brick as a building material, and the continuing importance of traditional fenestration patterns, are clearly shown in figure 10.12.

The architectural historicism of the 1940s and 1950s in Lleida is distinct in form and concept from that common in post-modernism (cf. Jencks and Chaitkin 1982). These earlier historicist developments were, in part, an extension of local vernacular architecture (figure 10.13).

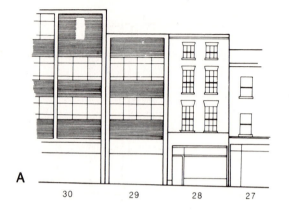

A

30   29   28   27

B

C

Figure 10.9 Proposals for the redevelopment of 28–9 High Street, Worcester (redrawn from planning applications).

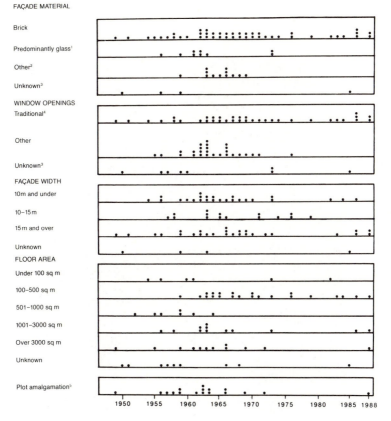

Figure 10.10 Characteristics of new buildings in the commercial core of
Worcester.

Notes: 1 Includes all types of wall where glass is the predominant material
2 Includes a variety of structural and facing materials such as stone, artificial stone, concrete and ceramic tiles
3 Normally, single-storey buildings with large, plate-glass display windows
4 Window openings of traditional proportions
5 Series incomplete owing to missing data

They were comparable to the neo-Georgian developments in Worcester in their impact on the urban landscape: neither style implied a radical break with tradition, as the modern style did. Glass façades and long balconies were used far more in modern-styled developments in Lleida than they were in Worcester (the balconies in Lleida being a vernacular

Figure 10.11 Example of a building in post-modern vernacular style, Worcester, built in the 1980s (photograph by the author, 1988).

tradition), and their quantitative impact was highly significant since the building regulations in force at that time, together with some flexibility on the part of the Council, allowed multi-storey development (figure 10.14).

In strong contrast to the largely vernacular post-modernism prevalent in Worcester (figure 10.11), and indeed in the other British towns that have been studied in a similar manner (Larkham and Freeman 1988a: figure 2), recent developments in Lleida have included stylistic developments of modernism together with other forms that, under the classification of Jencks and Chaitkin (1982), must be considered as

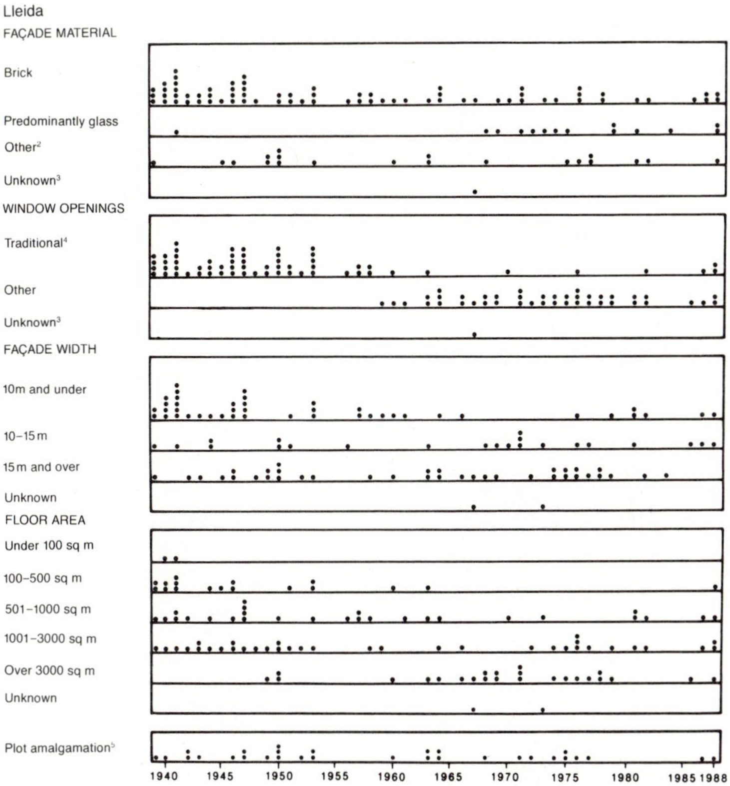

Figure 10.12 Characteristics of new buildings in the commercial core of Lleida.
Notes: 1 Includes all types of wall where glass is the predominant material, including glass screens
2 Includes a variety of structural and facing materials such as stone, artificial stone, concrete and ceramic tiles
3 Normally, single-storey buildings with large, plate-glass display windows
4 Window openings of traditional proportions, including balconies and enclosed balconies
5 Series incomplete owing to missing data

post-modern (figure 10.15). These differing forms of the post-modern style have had markedly different impacts on the urban landscapes of the two commercial cores.

Alterations to historical buildings have been abundant in Worcester and rare in Lleida (figure 10.7). Of particular importance, however, is

Figure 10.13 Architectural historicism as a development of vernacular style: Lleida (photograph by the author, 1988).

the type of alteration: those in Lleida have been more significant in terms of the urban landscape, and there appears to have been a more 'assertive' attitude on the part of the agents involved. The three examples of the immediate post-war period show a lack of sensitivity on the part of the developer to the character and appearance of historic buildings, and the lack of a policy on historical preservation on the part of the Council that might have prevented these developments. Figure 10.16 shows the example of post-war residential development over-

Figure 10.14 Example of how Lleida building regulations allowed tall modern-styled buildings to dominate their neighbours (photograph by the author, 1988).

whelming the last remaining medieval gateway into the old city core. During the 1960s and early 1970s, as in Worcester, an emphasis on the contrast between old and new architectural styles prevailed in developments affecting historic buildings. The latest changes, in the 1980s, show greater respect for the architectural styles of existing buildings, although the substantial remodelling of the Plaça de Sant Joan remains controversial (figure 10.17).

The urban landscapes of the two commercial cores must be understood as a result of a series of developments and their impacts on the existing urban fabric. In Worcester, a line of development may be

287

Figure 10.15 Example of a building in a post-modern style in Lleida, showing less historicism (redrawn from a planning proposal, 1980s).

observed, beginning with an important body of Georgian and vernacular buildings, continuing with Victorian styles of the mid- and late nineteenth century, neo-Georgian styles in the inter-war period of the twentieth century and part of the post-war period, and finally vernacular post-modern. The neo-Georgian style was the alternative suggested by local planning officers in the face of modern-styled development proposals. This was understandable in the inter-war period, when neo-Georgian was an accepted style, particularly for public buildings (cf. Whitehand 1984), and in a city with a large body of Georgian buildings (recognized by planners, for example Minoprio and Spencely 1946), but unusual in the extent of its continuing use in the post-war period. Yet use of this style allowed new buildings to blend with the old. The use of vernacular post-modern allows some measure of blending, often by the use of analogous forms, but post-modern styles do not, on the whole,

Figure 10.16 Example of early post-war development overpowering one of Lleida's historic gates (redrawn from a planning proposal).

disguise the fact that buildings are new. The continuity of style in Worcester was broken by the emergence of the modern/international style, and by the substantial alterations to the historical plot pattern enforced by the planning policies of the immediate post-war years. Even so, the modern style in Worcester was moderated, to a considerable extent, by development control policies; and the use of modern styling in the 1960s can, perhaps, best be seen as a following of national (and international) fashions (Larkham and Freeman 1988b), an exception in the evolution of architectural style within the city.

In Lleida, in contrast, the continuity in styles that is detectable derives from the vernacular tradition of rural building, persisting in this urban context until the post-war period. This is paralleled by a series of historicist styles between the last decades of the nineteenth century and the 1950s. The break of continuity is clear, with the emergence of the modern/international style from the 1960s onwards. Building regulations and Council permissions increased the impact of these modern buildings on the landscape of traditional architectural styles, building heights and masses and plot sizes. Furthermore, the non-contextual,

289

Figure 10.17 View of the remodelled Plaça de Sant Joan, Lleida, showing steps and walkway to the Seu Antigua (photograph by P. J. Larkjham, 1986).

post-modern styles used here (figure 10.14) show a greater similarity to the modern urban landscape than to the traditional landscape. The lack of development control and the weak development of regulations concerned with aesthetics are central to this aspect of the development of Lleida's urban landscape.

URBAN AGENTS AND ARCHITECTURAL STYLES

Previous research in Britain has related various characteristics of the agents of change, particularly those of developers and architects, to the architectural style and type of building produced (Whitehand 1984; Freeman 1986). The process of innovation diffusion down the urban hierarchy has been suggested, a diffusion in which novel styles were introduced by large developers and architects based outside the city, and then adopted and further diffused by local agents. In Worcester the process of diffusion of modern and post-modern styles can be explained in a similar way. These styles were adopted first by non-local architects and developers, and diffused to local agents. Many of the non-local agents involved were based in London. National commercial and financial chains tended to spread the new styles to smaller businesses, institutions and speculative developers. The large companies were the first to leave traces of the new styles and thus to transform, in a more radical manner, large urban areas. On the other hand, the neo-Georgian developments of the 1960s must be seen as a strategy on the part of local planning officers seeking to retain a local style and thus preserve the historical urban environment.

In Lleida, the characteristics of diffusion of the modern and postmodern styles are similar to those of Worcester, but the processes are not so clear. Non-local developers and architects, and financial chains and large companies were involved in the introduction of new architectural styles, and it was they who were responsible for the buildings in the modern/international style that had such a great impact upon the urban landscape. But it would seem that there was a very fast adoption of innovations by these local agents.

An interesting variation on this is the case of the architects working with post-modern styles. These, in the case of developments in Lleida, all had local addresses. The initiators were mostly based in offices in Barcelona and Lleida – the two cities are closely linked in the economy of the Catalan region – and, in fact, the architects were working from local offices of firms whose principal office was, in most cases, in Barcelona. A further point is the difference in the number of architects at work in the two city centres. In central Worcester, many different architects were employed, with only two practices designing more than one building. In central Lleida, in contrast, the majority of buildings constructed between the Civil War and the late 1960s were designed by only six local architects, using different styles ranging from historicism and vernacular to modern. Only in the 1970s and 1980s has the number of architects working on developments in Lleida's commercial core increased significantly.

DIFFERENCES BETWEEN THE CITY CENTRES

A first group of differences between the two commercial cores studied reflects their cultural diversity. Here there are two fundamental aspects. First, there is the combination of residential and non-residential land uses in Spanish town centres, as opposed to the almost wholly non-residential uses common in British town centres. In this respect, the role of planning policies in the respective city centres is important, with the restriction until recently on land uses in Worcester contrasting with the policies supporting speculative housebuilders in Spain until the mid-

1970s (see, for example, Naylon 1986; Vilagrasa 1987).

Second, there is the persistence of small-scale commercial enterprises in Lleida, in contrast to the enormous impact made by companies of regional or national size in Worcester. Thus corporate capitalism in Britain involves a relatively high concentration of financial, commercial and building activities. This was a condition largely absent from Spain during the study period, reflecting that country's peripheral role in the development of Western capitalism generally.

A second set of differences relates to the distinct municipal urban managerial traditions in Great Britain and Spain. Great Britain has developed a very complex system of development control, particularly for dealing with aesthetic matters (Punter 1986b, 1987). This system gives great freedom to the planning officers of the Local Planning Authority when recommending the granting, or refusal, of planning applications on technical or aesthetic grounds. And although the actual decisions are made by an elected Planning Committee, only in a small minority of cases does the Committee deviate from the planning officer's recommendations (Fleming and Short 1984). This system has meant that the preservation of the historical urban landscape, an objective embodied in formal planning terms in the 1967 Civic Amenities Act but recognized, in Worcester's case, by the first post-war planning proposals (Minoprio and Spencely 1946), has largely been maintained. The objectives of preservation include the protection of valued buildings, and have influenced the fact that a number of modern-styled development proposals have been adapted to take account of the architectural styles of existing buildings. Controls on design were strengthened by the designation of much of the city centre as a Conservation Area, since aesthetic control operates in such areas rather more effectively than in non-designated areas (Punter 1986a). In contrast, Spanish planning tradition places the emphasis firmly on quantitative standards, particularly building heights, volumes and widths, and only on rare occasions are aesthetic standards stipulated. Moreover, development control has been practically non-existent in Spain during this study period. A major effect of these two distinct national approaches has been the creation of two different urban landscapes.

In Worcester, even in the case of some modern buildings, there has been a requirement that materials be used that are similar to those already common in the area, for example, brick or stucco, and that certain features of design be adopted, for example the substitution of traditional fenestration patterns for proposed modern horizontal windows or glass façades. But many of these decisions are subjective, and depend on the aesthetic tastes of the local planning officers, not on any objective and well-informed vision that stipulates what the urban form should be. Even so, some compatibility with existing architectural styles can be seen in a number of developments. Stylistically, development controls and constraints have been formulated according to the aesthetic taste of each period, hence the continuity of neo-Georgian styles into the 1960s. Furthermore, the vernacular post-modern usage of new materials and construction techniques still produces forms similar to those of local vernacular buildings.

In Lleida, however, planning practices have allowed a great freedom of architectural styles. This permissiveness, which reached a peak in the 1960s and 1970s, in combination with increases in building heights and volumes, has resulted in radical changes in the urban landscape. Furthermore, the lack of any clear policy to protect historical buildings has resulted in the destruction and replacement of buildings that would now be considered worthy of preservation. Beginning with the General Plan of 1979, the Listed Building Plan of 1983 and the Special Plan for the Historical Centre of 1985, conservation policies have begun to establish basic aesthetic directives to constrain building in the city centre. Despite this, the standards that have been laid down have not been enforced to any great extent, and there has been a lack of real control over development. In practice, the municipal planners have not been very critical of the projects proposed to them. There has been a failure to ensure continuity between the existing urban landscape and new development proposals. For example, in post-modern projects the adoption of popular local styles is frequently discussed, but a clear idea of what this may mean is manifestly missing. The proposed buildings are of non-local design and are merely camouflaged as 'vernacular'; they do not fit in with their surroundings in texture, colour or form. There is a clear parallel here with the rise of a nationwide 'pseudo-vernacular' in Britain, in which there is a failure to use truly local styles and materials (Larkham 1988b).

CONCLUSION

Accepting the difficulty of transporting the British planning system, with its few set standards and reliance upon subjective aesthetic criteria, into Spain, with its many standards and regulations, and accepting also that the perception by British people of the effects of their conservation planning system is not as favourable as that presented here (Fergusson 1973; Aldous 1975), an effort should be made to integrate the necessary economic stimulation of business and commerce in the two centres within a framework in which new development would respect the historical urban environment. In the first place, non-quantitative regulations should be developed to establish acceptable criteria for 'historical' and 'aesthetic' standards in new developments. Secondly, development control should be understood as a key to a policy for urban conservation. A dialogue should be developed between developers, architects and town planners in which constructive ideas are developed that are both worthwhile for private investment and, at the same time, favourable to the conservation of the urban historical heritage.

ACKNOWLEDGEMENTS

This research was funded by the Comissió Interdepartmental de Recerca i Innovació Tecnològica de la Generalitat de Catalunya (CIRIT). A more detailed report of the Worcester part of the study is published as Vilagrasa (1990). This chapter was prepared for English publication by P. J. Larkham and J. W. R. Whitehand.

REFERENCES

Aldous, T. (1975) *Goodbye, Britain?*, London: Sidgwick and Jackson.

Alió, M. A. (1987) 'Els expedients d'obres particulars com a eina d'anàlisi del procés urbà. Vilafranca del Penedès, 1845–1945', in Morell, R. and Vilagrasa, J. (eds) *Les ciutats petites i mitjanes a Catalunya: evolució recent i problemàtica actual*, Barcelona: Institut Cartogràfic de Catalunya.

Fergusson, A. (1973) *The Sack of Bath*, Salisbury: Compton Russell.

Fleming, S. C. and Short, J. R. (1984) 'Committee rules OK? An examination of planning committee action on officer recommendation', *Environment and Planning A* 16: 965–73.

Frampton, K. (1980) *Modern Architecture: A Critical History*, London: Thames and Hudson.

Freeman, M. (1986) *Town-centre Redevelopment: Architectural Styles and the Roles of Developers and Architects*, Occasional Publication 20, Birmingham: Department of Geography, University of Birmingham.

Gallart, J., Junyent, E., Pérez, A. and Rafel, N. (1985) *L'Arqueologia a la ciutat de Lleida 1975–1985*, Quaderns de divulgació ciutadana 5, Lleida: Col·lecció: La Banqueta.

Gomez Mendoza, A. (1986) 'La industria de la construcción residencial en Madrid 1820/1935', *Moneda y Crédito* 117: 53–81.

Grup d'Estudis Urbans (1987) 'Promoció immobiliària i característiques de l'edificació en les petites ciutats de l'entorn de Lleida: Alcarràs, Les Borges Blanques i Mollerussa', in Morell, R. and Vilagrasa, J. (eds) *Les ciutats petites i mitjanes a Catalunya: evolució recent i problemàtica actual*, Barcelona: Institut Cartogràfic de Catalunya.

Gutkind, E. A. (1967) *Urban Development in Southern Europe: Spain and Portugal* (volume III of the *International History of City Development*), New York: Free Press.

Jencks, C. A. and Chaitkin, W. (1982) *Current Architecture*, London: Academy Editions.

Larkham, P. J. (1986) *The Agents of Urban Change*, Occasional Publication 21, Birmingham: Department of Geography, University of Birmingham.

Larkham, P. J. (1988a) 'Changing conservation areas in the English Midlands: evidence from local planning records', *Urban Geography* 9, 5: 445–65.

Larkham, P. J. (1988b) 'Agents and types of change in the conserved townscape', *Transactions of the Institute of British Geographers* NS 13, 2: 148–64.

Larkham, P. J. and Freeman, M. (1988a) 'A re-examination of reasons for using building styles', *Local Historian* 18, 4: 183–6.

Larkham, P. J. and Freeman, M. (1988b) 'Twentieth-century British commercial architecture', *Journal of Cultural Geography* 9, 1: 1–16.

Larrosa i Padró, M. (1987) 'La construcció i l'habitage a través de les llicènces d'obres: Sabadell, 1900–1938', in Morell, R. and Vilagrasa, J. (eds) *Les ciutats petites i mitjanes a Catalunya: evolució recent i problemàtica actual*, Barcelona: Institut Cartogràfic de Catalunya.

Lladonosa i Pujol, J. (1983) *Els carrers i places de la Lleida actual amb més pes històric*, Quaderns de divulgació ciutadana 2, Lleida: Col·lecció: La Banqueta.

Minoprio, A. and Spencely, H. (1946) *Worcester Plan: An outline Development Plan for Worcester*, Worcester: City Council Reconstruction and Development Committee.

Naylon, J. (1986) 'Urban growth under an authoritarian regime: Spain 1939–1975: the case of Madrid', *Iberian Studies* 15, 1–2: 3–15.

Punter, J. V. (1986a) 'The contradictions of aesthetic control under the Conservatives', *Planning Practice and Research* 1: 8–13.

Punter, J. V. (1986b) 'The history of aesthetic control, 1909–53', *Town Planning Review* 57, 4: 351–81.

Punter, J. V. (1987) 'The history of aesthetic control, 1953–85', *Town Planning Review* 58, 1: 29–62.

Vilagrasa, J. (1984) 'Creixement urbà i producció de l'espai a Lleida (1940–1980)', *Documents d'Anàlisi Geogràfica* 5: 97–138.

Vilagrasa, J. (1987) 'Política de l'habitatge i promoció privada a Lleida (1940–1980)', *Revista Catalana de Geografia* II, 5: 33–50.

Vilagrasa, J. (1990) *Centre històric i activitat comercial: Worcester 1947–1988*, Espai/Temps, Lleida: Estudi General de Lleida.

Vilagrasa, J. (forthcoming) 'Les llicencies d'obres com a font d'estudi de l'urba: una analisi comparativa entre Gran Bretanya i Espanya', in Hernando, A. and Panareda, J. M. (eds) *Homenatge al Professor Lluís Casassas i Simó*, Barcelona: Universitat de Barcelona.

Whitehand, J. W. R. (1984) *Rebuilding Town Centres: Developers, Architects and Styles*, Occasional Publication 19, Birmingham: Department of Geography, University of Birmingham.

Whitehand, J. W. R. and Whitehand, S. M. (1983) 'The study of physical change in town centres: study procedures and types of change', *Transactions of the Institute of British Geographers* NS 8, 4: 483–507.

Whitehand, J. W. R. and Whitehand, S. M. (1984) 'The physical fabric of town centres: the agents of change', *Transactions of the Institute of British Geographers* NS 9, 2: 23–47.

CONCLUSION

11

URBAN LANDSCAPE RESEARCH: ACHIEVEMENTS AND PROSPECTUS

J. W. R. Whitehand and P. J. Larkham

CHANGES IN URBAN MORPHOLOGY

It is remarkable that in a review of research on urban change in the UK prepared as recently as the late 1980s, Fielding and Halford (1990: 10) should draw attention to how little research had been undertaken on the physical form of cities. Fortunately, as the foregoing chapters have shown, this would be an unduly pessimistic view of the state of research in urban morphology in the early 1990s. The urban landscape is suddenly attracting a great deal of interest among scholars in several different disciplines – in geography, planning and planning history, urban design, architecture and urban history. Among the manifestations of this interest are not only the conference upon which this volume is based but also the publication twice yearly since 1987 of the international *Urban Morphology Newsletter*, the formation within the international research network URBINNO of a working group on land use, urban morphology and the physical environment (Curdes and Montanari 1990), the publication of a *Glossary of Urban Form* (Larkham and Jones 1991), a marked increase in the rate of production of scholarly papers on aspects of urban form (Whitehand 1992), and the publication of several books providing conspectuses of the urban land-scape (Slater 1990, Vance 1990, Whitehand 1991).

The contents of this volume are in some respects an index of this change. That they reflect the interdisciplinary and international coming-together that was referred to in chapter 1 is especially evident. But they also reflect other changes, including changes in the way in which research in urban morphology is being conducted. Of particular note is the increase in the amount of direct collaboration between researchers. For example, town-plan analysis had until recently been essentially a task for the individual scholar. Most notably, Conzen's writings were the product of individual scholarship, in most cases even to the extent of the author's preparing his own drawings for publication. Now joint and group projects are developing, exemplified in this volume by Baker and Slater's study of morphological regions. Conzen's much earlier lone scholarship having provided the basis for this field, others are at last beginning to co-ordinate their efforts. Similar co-operation is to some extent evident among medieval archaeologists. It is most evident in studies of twentieth-century urban landscapes in Great Britain undertaken by the Urban Morphology Research Group at Birmingham. The study of low-density residential suburbs, described in chapter 9, is an example. In such work, large amounts of information are being collected to the same format by several researchers. This is effectively the only practical way of collecting the large quantities of information necessary if the conclusions drawn are demonstrably to be of more than local significance. This is not to suggest that this type of research readily lends itself to team work. A major investment of time is necessary in standardizing definitions and procedures if the satisfactory pooling of data and reliable comparisons of data collected by different researchers are to be achieved. Vilagrasa's comparative study of historic city centres in England and Spain, undertaken in collaboration with the Urban Morphology Research Group, is an illustration in this volume of the extension to an international level of attempts to standardize such comparisons. In such cases the problems of compiling truly comparable data are made worse by the national differences in the nature of the administrative records that constitute such an important source. Nevertheless, with careful preparation of procedures, it is still possible to make meaningful comparisons.

The problems of establishing standard definitions in urban morphology and the fact that much of the information on urban form is not readily converted into 'data' has retarded the large-scale use of computers in storing and processing information. However, again, careful preparation can provide the basis for remarkable computer-aided investigations, as Holdsworth has shown in the three-dimensional reconstruction of the development of Manhattan in this volume. There would seem to be wide scope for further computer-aided research of this type. And fortunately much of the necessary technical capability is already available. However, it is in this field that the contact between research perspectives is weakest. The Conzenian tradition has developed largely in isolation from computer-aided analyses of urban form and vice versa.

COMPUTERS AND URBAN MORPHOLOGY

The dividing line between urban morphogenetics and essentially descriptive, computerized analyses of urban form is probably the least crossed of the numerous boundaries that are characteristic of urban morphology. That several of the types of investigations described in this volume could benefit from the breaking down of this boundary is suggested even by a cursory review of recent computer-aided research on urban form.

This research falls into three principal categories. The first has to do with the three-dimensional form of urban areas, and is particularly concerned with aiding geometrical composition so that proposals for new forms, or the adaptation of existing forms, can be evaluated both visually and in terms of functional efficiency. The second is primarily concerned with the analysis of physical structures, especially individual structures such as dwelling houses, viewed in two dimensions. In the third, the accent is upon urban areas, or land-use parcels, as physical configurations, especially as represented cartographically.

Developments in both the software and the hardware for computer graphics have now reached the stage where realistic simulations of urban landscapes are possible. One of the most striking applications is in the reconstruction of former urban landscapes; this has been demonstrated by Cicconetti *et al.* (1991) in their reconstruction of part of central Genoa as it was in the late fifteenth century. They show that by combining detailed historical and archaeological research with advanced computer graphics it is possible to simulate a walk through part of the late-medieval city. Such technical advances can also be used to visualize the impact on existing urban landscapes of proposed new buildings or modifications to existing buildings (Grant 1991), with obvious value in the types of urban-landscape management called for in chapter 9.

Much of the quantitative, or more precisely geometrical, analysis of buildings viewed two-dimensionally relates to attempts to develop a science of architectural form. Steadman *et al.* (1991) argue that it is only by developing theories that explain why certain plans and built forms rather than others occur in practice that scientific generalizations can be made about the relationships of built forms to the functions they fulfil. Since the large majority of rooms in domestic buildings are essentially rectangular in plan, the plans of most dwellings can be modelled mathematically as 'packings' of rectangles within rectangular boundaries. The enumeration of all possible packings by computer methods provides a complete 'map' of the theoretical space of geometrical possibilities within which floor plans can occur. 'The boundaries and topography of this space are fixed and immutable for all time, and all architects, past, present and future, have no choice but to work within them' (Steadman *et al.* 1991: 88). Within the boundaries there are, of course, further effective constraints on 'geometrical possibility' imposed by, for example, technological factors and legal requirements, and the need for daylighting and access. These constraints change historically and account for major reductions in practice in the numbers of plans that considerations of geometrical close-packing alone would permit. Comparison of theoretical possibilities with dwelling plans existing in reality serves to highlight how, even in the case of simple plans, social needs may override material ones.

Brown and Steadman (1987) believe that the process that they illustrate, namely one of exhaustive plan generation under constraints, offers a tool that is useful in both architectural design and in helping to fill gaps in the historical and archaeological record. As they recognize, however, their procedure involves a certain circularity of explanation: a set of constraints is inferred by reference to the physical characteristics of actual building plans, which is then used to account for those physical characteristics (Brown and Steadman 1987: 436). Such problems notwithstanding, applications of the generation of building plans by computer are now being widely explored, but for the present they remain far removed from the types of plan analysis discussed by Baker and Slater in this volume.

The application of computers to the analysis of the shapes of urban areas has proceeded for the most part separately from analyses of both floor plans and three-dimensional urban form. Much of the concern with the shapes of urban areas has focused on their degree of irregularity. It has become evident in recent years that fractal geometry provides an appropriate means of measuring many types of irregular form that had previously resisted scientific classification. Batty and Longley (1987, 1988) have been at the forefront of its application to urban areas. They conclude from their analysis of the urban boundary of Cardiff between 1886 and 1922 that the traditional image of urban growth becoming more irregular as tentacles of development occur along railway lines is not borne out. They point out, however, that the temptation to explain this as an effect of increasing controls over the environment should be resisted because there is considerable variation in the results produced by different methods (Batty and Longley 1987). They also conclude that there is tentative evidence that parcels of land used for residential purposes and open space are more irregular than parcels of land used for commercial/industrial purposes, education and transportation. Again, however, there remains considerable uncertainty about the processes in operation (Batty and Longley 1988). Although Batty (1991) emphasizes the need for better measurements of urban development and density, there is an even greater need to bring together this research and studies of the activities responsible for the form of urban development, such as appear in several chapters in this volume.

HUMANISM AND THE URBAN LANDSCAPE

Despite this need to link the various types of research in urban morphology, there are certain investigative styles that, because of their intrinsic character, seem more likely to follow essentially separate scholarly paths. They are largely humanistic in character, concerned with interpretation rather than analysis. The two essays in this volume that come closest to this *œuvre* are that by Friedman on palaces and the street in late-medieval and Renaissance Italy, and that by Knox on post-modern American suburbia. But it is important to reflect on the contents of these two essays, and indeed the other chapters, against a wider spectrum of work of a humanistic disposition.

Particularly noteworthy among such work is a growing concern with the social significance of urban forms, a line of investigation that Knox has pursued in this volume. In particular, the symbolic qualities of urban landscapes have attracted interest. But seeking to uncover the meanings that human beings ascribe to urban landscapes is a delicate task that is fraught with difficulty. A notable development, though its utility remains questionable, has been the recourse that scholars have had to the methods of linguistics and semiology. This has led to urban landscapes being viewed as 'texts' to be interpreted. Such interpretations are far removed from the idea of landscape as a mathematical or statistical construct. Instead the focus of interest is often the ideological basis of creations in the landscape. It has been suggested by Duncan and Duncan (1988) that the most fruitful development of this intellectual perspective may take the form of the integration of literary theory and social theory. They argue that literary theory provides ways of examining the text-like quality of landscapes, and of understanding them as transformations of ideologies. It also offers ideas on reading and authorship that can be adapted to explain how landscapes are incorporated into social processes. But there is, in addition, a need to take account of social organization. This can be done by introducing the

303

notion of 'textual communities', social groups that, broadly speaking, share a common reading of a text.

Despite the amount of discussion that has appeared on landscapes as ideological constructions, the amount of empirical work is small. One of the more convincing studies is that by Ley and Olds (1988) on international expositions. They consider the view that such spectacles are examples of the manipulation of the mass of the population by an élite. They conclude that imputations of hegemony and social control in existing studies contain minimal evidence of the actual perceptions of the public. Their own analysis of the 1986 Exposition in Vancouver and of the perceptions of visitors to that spectacle concluded that the cultural dupes posited by mass-culture theorists were less visible on the ground than in speculation on the subject. But, unfortunately, the contributions to this volume by Friedman and Knox notwithstanding, it is unconfirmed speculation that is in the ascendant among studies that adopt this perspective on the urban landscape.

Among scholars adopting a humanistic approach to the urban landscape, considerable interest has surrounded the nature and manifestations of post-modernism. The notion of post-modernism has been regarded by some researchers, including some of a primarily social science disposition, as sufficiently fundamental to justify speculation about links between post-modernism and economic changes. Indeed, at the beginning of this volume attention was drawn to the fact that an important part of the stimulus for the growth of interest in the urban landscape is that the appearance of new landscape features and changing spatial distributions since the mid-1970s has been seen as a manifestation of fundamental economic transformations. These suggested economic changes, associated with reduced emphasis on mass production, have prompted questions about the validity of existing urban theories. It has been suggested that post-modern urban landscapes do not conform to the sectors and zones that were recurrent elements in previous debates about the internal structure of cities (Knox 1991: 203). However, the examination of a long span of history in this volume does not provide grounds for regarding the onset of post-modernism as providing changes to the urban landscape that are notably more fundamental than those that were characteristic of the onset of previous morphological periods. Within Western countries in general and the UK in particular, the years following the First World War brought new urban landscapes, particularly new residential landscapes, that arguably differed more fundamentally from their predecessors than post-modern landscapes have differed from the landscapes created in the three decades following the Second World War. Indeed one of the key features of many post-modern landscapes is the fact that they differ from landscapes created in the 1950s and 1960s in ways that are essentially cosmetic. It is true that residential high-rise building became unfashionable in the 1970s in most Western countries, but the principal change in the appearance of low-rise residential building has been in the greater use of external decoration and the greater variety of house type contained within individual streets. The historicist elements characteristic of much of architectural post-modernism are, in many cases, merely applied to building elevations. These are superficial changes compared with those that heralded some earlier morphological periods. For example, the introduction of Renaissance architectural theory, as Friedman has shown in this volume, reintroduced the concept of the façade and entailed a transformation in the way in which the urban landscape was viewed.

THE MANAGEMENT OF URBAN LANDSCAPE CHANGE

Nearly all of the research that has been presented in this volume has implications for the management of urban landscapes. An issue that has become of increasingly critical significance to any such management in recent years is the relationship between new development and existing urban landscapes. Research on urban landscape conservation is one aspect of a heightened concern for buildings inherited from past periods. A significant part of this concern stems from dissatisfaction with the physical forms, especially the buildings, that have been produced in the twentieth century. Jacobs and Appleyard (1987) attribute the problem in large part to the fact that the urban environment has become increasingly controlled by large-scale developers and public bodies. The large areas and large complexes that they create make people feel irrelevant. People have less sense of control over their homes, neighbourhoods and cities than when they lived in slower-growing, locally focused communities. For Jacobs and Appleyard it is through involvement in the creation and management of their city that people are most likely to identify with it.

Other criticisms have been directed at development control systems, for example in Perth, Australia (Yiftachel 1989: 63–4) and Chelmsford, England (Hall 1990). Hall contends that a major problem in Great Britain is the failure of planning authorities to formulate explicit objectives for the design of different parts of urban areas. In similar vein,

Whitehand (1991) concludes that, at the scale of the British streetscape, governmental influence is often less today than that of major nineteenth-century estate owners exercising control over the development of their land. Planning authorities are largely reacting to proposals whose formulation and initiation are outside their control. Unlike major nineteenth-century landowners, local planning authorities in Great Britain plan specific landscapes only rarely. In response to specific proposals they seldom suggest, except in the most general terms, the type of landscape that they regard as appropriate. They state what is unacceptable according to their rules and procedures, but their creative role is in general very limited, as has been shown in chapter 9. And this role became even smaller under the Conservative administration of the 1980s (Punter 1986a). Subtle qualities of a landscape, such as the *genius loci*, pale into insignificance as influences upon development control decisions in comparison with measurements of building density and the dimensions and geometry of highways.

This view is to some extent consistent with the conclusion of Punter's (1985) study of office building in the commercial core of Reading. He points out that aesthetic considerations are the first to be sacrificed in the cause of 'speed and efficiency' in decision-making by clients, developers, architects and planners. Developers have had a large measure of freedom and have felt compelled by the requirements of letting and funding, more than by planning control, to keep within the mainstream of architectural fashion. The major pressures on development control planning officers are for speed and efficiency in making decisions, measured crudely in terms of weeks elapsed from the submission of an application (Larkham 1990).

Punter (1985, 1990) is exceptional in the detail of his exposition of how aesthetic control operates, or fails to operate. More common are studies of planning legislation and plans, as distinct from actual developments in the landscape. Much of the literature on conservation has been of this type. But conservation policies frequently lack effective means of implementation. Even more important, as M. R. G. Conzen (1975) pointed out long ago, they lack a theoretical basis – a theory of urban landscape management that can give direction and coherence to the way in which conservation problems are tackled. The few approaches to conservation in Great Britain that have a theoretical content (Briggs 1975; Faulkner 1978) have still not found their way into the mainstream planning literature. In an attempt to fill this theoretical vacuum, Kropf (1986) has re-examined the approaches of M. R. G. Conzen and Caniggia to the management of urban landscapes. The positions reached by these two scholars have much in common. For Kropf they afford a means of discovering a theoretical structure that underlies the relationship between the historico-geographical explanation of the development of urban forms and the prescription of urban design.

An essential part of the thinking of Caniggia, as of that of M. R. G. Conzen, is the view that the intelligibility of the city depends upon its history. In formulating a basis upon which urban landscapes can be managed, it is a short step from this fundamental belief to regarding urban forms as a source of accumulated experience, and from there to utilizing this experience as the basis for prescribing change. Possible solutions may be 'read' from the existing landscape, but they must be assessed to ensure that they are appropriate to new problems (Whitehand 1991). It is particularly important that the significance, including the historical significance, of urban landscapes for those experiencing them is understood. This includes the emotions, sometimes disagreeable emotions, that landscapes evince.

As is apparent from the detailed case studies described in chapter 9 of this volume, the perceptions of urban landscapes by academic theorists need to be related to those of the general public, if they are to contribute effectively to planning practice. The examination by Groat (1986) of the views of members of the public in the American Mid-West on the compatibility of new buildings with the existing urban landscapes into which they had been inserted is enlightening in this respect. Of particular interest is her examination of public reactions to the view that 'architecture should be treated as an historic document'. Her conclusion was that this view has little support. One inference to be drawn would seem to be that considerations of the historical associations or historical expressiveness of buildings are, to say the least, a less significant part of the landscape appreciation of those interviewed than other considerations. Of overriding importance for the general public, Groat concluded, was that new buildings should visually harmonize with those existing. This conclusion is consistent with the findings of Punter (1986b: 208). It would seem that for most people it is outward appearances, rather than historicity, that is crucial. Indeed, Bishop (1982) found that people who had mistakenly believed particular buildings to be genuinely Tudor were not at all disturbed to find that they were fakes. That the appreciation of an architectural style is independent of an awareness of its historical associations or expressiveness is, however, a much more questionable proposition. An appreciation of 'fakes' in the landscape will frequently be influenced by a knowledge of the cultural context of the originals that have been copied. For those who know anything about the history or historical associations of a style, an ahistorical appreciation is impossible (Whitehand 1991: 208).

THE INTEGRATION OF RESEARCH PERSPECTIVES

That research on the urban landscape is proceeding with vigour on a number of fronts is clear from the essays that have been brought together in this volume. The distinctions between fronts are often inter-disciplinary, but not exclusively so. For example, within geography alone three distinct types of research are apparent – the computer-aided analysis and representation of urban physical form, urban morpho-genetics, and the exploration of the social significance of urban land-scapes. Fortunately, there is increasing evidence that some of the bound-aries between the types of research that have been recognized are being bridged. Furthermore, attempts are being made to develop links between the types of research described in this volume and other fields of knowledge. The most ambitious attempted linkage is between, on the one hand, the cultural changes (most obviously those manifested in changes of architectural style) that have been interpreted by various authors as a shift from modernism to post-modernism, and on the other hand, shifts in the character of capitalism (Harvey 1989). Meanwhile, *within* urban morphology there are attempts to integrate hitherto separate strands of research. One such attempted link, already referred to, between urban morphogenetics (especially the ideas of M. R. G. Conzen) and Italian architecture, has the double impediment of language barrier and interdisciplinary barrier to overcome. Concerned with developing a theoretical basis for urban-landscape management, its progress may ultimately benefit from the forging of links with hitherto largely unconnected developments in computer-aided design.

That language barriers are beginning to crumble is particularly evident within geographical urban morphology, most notably in the spread of the fringe-belt concept (Whitehand 1988; von der Dollen 1990; Vilagrasa 1990). And in addition to disciplinary boundaries being overcome, it is remarkable how research resembling that in the Con-zenian school has occasionally been undertaken quite independently (Moudon 1986). But there is considerable scope for the further inte-gration of hitherto largely separate strands of research. The strangely separate existences and perspectives of architectural history and urban history have been discussed by Arnold (1990). And as Knox (1987) has noted, studies of architects and architecture have on the whole taken remarkably little account of the economic and social context within which architects work, a consequence being that theories about architecture remain weakly developed. The problem is partly the wider one of the division between the viewpoint of the social sciences and that of the humanities. This is a division that creates particular difficulty for urban designers seeking to influence planning practice, since the discipline of planning, especially in America, is embedded within the social sciences rather than within the design professions (Stevens 1990).

The differences between the literatures in which studies of urban form occur and the problems of tunnel vision that have arisen are in part a reflection of the diverse nature of urban form itself. It is a notable feature of the early 1990s, therefore, that some of the most stimulating of the debates concerning the urban landscape are precisely those being carried out at the interfaces of different intellectual perspectives. Among the issues near the centre of these debates, two stand out (Whitehand 1992).

First, there is the relationship between the appearance of post-modern urban landscapes and the changing character of capitalism. The empirical work necessary to determine whether there is indeed a relationship and, if it exists, to ascertain its nature, has scarcely begun. The majority of discussion of post-modern landscapes is based on little more than anecdote. Spectacular post-modern landmarks have attracted attention, but the dominant characteristics of change in the great majority of urban landscapes since the mid-1970s have been the greatly increased incidence of pseudo-vernacular styles and the heightened concern for existing physical forms. The constraining influence of the physical framework of streets, plots and buildings created over many generations is, in some respects, acting in concert with a reaction against the Modern Movement. In assessing the extent to which post-modern landscapes can be reconciled with existing urban theories it is necessary to establish the attributes of the new landscapes with which these theories are required to cope. Much of this groundwork remains to be done.

Secondly, the creation of urban landscapes in the 1990s, many of them adaptations of existing landscapes, lacks a well-articulated planning theory. Research on this subject is proceeding more slowly than on almost any other dealt with in this volume, and there is a large gap between it and both planning practice and the views of the general public. Punter (1990) has shown how important it is for future urban landscapes that the lessons of planning history are absorbed and he has spelt out the nature of those lessons. Harris (1988) has refocused attention on the indispensability of both scientific and humanistic approaches to planning. It is clear that in the early 1990s there is an increasing awareness of the need for greater integration, both at the general level with which Harris was concerned and in terms of the more specific issues that have been dealt with in individual chapters of this book. Sensitivity to the historical development of urban landscapes and an awareness of their social significance are increasingly apparent. The harnessing of these attributes, taking advantage, where appropriate, of increasingly sophisticated developments in the storage, analysis and presentation of landscape data, provides an attractive prospect for urban morphology in the 1990s.

ACKNOWLEDGEMENT

Part of this chapter is adapted from a paper by J. W. R. Whitehand in *Urban Studies* (1992).

REFERENCES

Arnold, J. L. (1990) 'Architectural history and urban history: a difficult marriage', *Journal of Urban History* 17: 70–8.

Batty, M. (1991) 'Generating urban forms from diffusive growth', *Environment and Planning A* 23: 511–44.

Batty, M. and Longley, P. A. (1987) 'Urban shapes as fractals', *Area* 19, 3: 215–21.

Batty, M. and Longley, P. A. (1988) 'The morphology of urban land use', *Environment and Planning B: Planning and Design* 15: 461–88.

Bishop, R. (1982) 'The perception and importance of time in architecture', unpublished PhD thesis, University of Surrey.

Briggs, A. (1975) 'The philosophy of conservation', *Royal Society of Arts Journal* 123: 685–95.

Brown, F. E. and Steadman, J. P. (1987) 'The analysis and interpretation of small house plans: some contemporary examples', *Environment and Planning B: Planning and Design* 14: 407–38.

Cicconetti, C., Gasparini, E., Mastretta, M. and Morten, E. (1991) 'A computer graphic reconstruction of the architectural structure of medieval Genoa', *Environment and Planning B: Planning and Design* 18: 25–31.

Conzen, M. R. G. (1975) 'Geography and townscape conservation', in Uhlig, H. and Lienau, C. (eds) *Anglo-German Symposium in Applied Geography, Giessen-Würzburg-München, 1973, Giessner Geographische Schriften*, 1975: 95–102.

Curdes, G. and Montanari, A. (1990) 'Review forum', *Town Planning Review* 61: 209–11.

Dollen, B. von der (1990) 'An historico-geographical perspective on urban fringe-belt phenomena', in Slater, T. R. (ed.) *The Built Form of Western Cities*, Leicester: Leicester University Press.

Duncan, J. and Duncan, N. (1988) '(Re)reading the landscape', *Environment and Planning D: Society and Space* 6: 117–26.

Faulkner, P. A. (1978) 'A philosophy for the preservation of our historic heritage' *Royal Society of Arts Journal* 126: 452–80.

Fielding, T. and Halford, S. (1990) *Patterns and Processes of Urban Change in the United Kingdom*, London: HMSO.

Grant, M. (1991) 'Issue: integrated software system for the urban environment', *Environment and Planning B: Planning and Design* 18: 33–8.

Groat, L. N. (1986) 'Contextual compatibility: a study of meaning in the urban environment', unpublished paper presented to the Annual Meeting of the Association of American Geographers, Minneapolis.

Hall, A. C. (1990) 'Generating urban design objectives for local areas: a methodology and case study application to Chelmsford, Essex', *Town Planning Review* 61: 287–309.

Harris, B. (1988) 'The emerging unity of science and humanism in planning', *Journal of the American Planning Institute* 54: 521–4.

Harvey, D. (1989) *The Condition of Postmodernity*, Oxford: Blackwell.

Jacobs, A. and Appleyard, D. (1987) 'Toward an urban design manifesto', *Journal of the American Planning Association* 53: 112–20.

Knox, P. L. (1987) 'The social production of the built environment: architects, architecture and the post-modern city', *Progress in Human Geography* 11: 354–78.

Knox, P. L. (1991) 'The restless urban landscape: economic and socio-cultural change and the transformation of metropolitan Washington DC', *Annals of the Association of American Geographers* 81, 2: 181–209.

Kropf, K. S. (1986) 'Urban morphology considered', unpublished MA thesis, Oxford Polytechnic.

Larkham, P. J. (1990) 'The concept of delay in development control', *Planning Outlook* 33, 2: 101–7.

Larkham, P. J. and Jones, A. N. (1991) *A Glossary of Urban Form*, Historical Geography Research Series 26, London: Institute of British Geographers.

Ley, D. and Olds, K. (1988) 'Landscape as spectacle: world's fairs and the culture of heroic consumption', *Environment and Planning D: Society and Space* 6: 191–212.

Moudon, A. V. (1986) *Built for Change: Neighborhood Architecture in San Francisco*, Cambridge, MA: MIT Press.

Punter, J. V. (1985) *Office Development in the Borough of Reading 1954–1984: A Case Study of the Role of Aesthetic Control within the Planning Process*, Working Papers in Land Management and Development Environmental Policy 6, Reading: University of Reading.

Punter, J. V. (1986a) 'The contradictions of aesthetic control under the Conservatives', *Planning Practice and Research* 1: 8–13.

Punter, J. [V.] (1986b) 'Aesthetic control within the development process: a case study', *Land Development Studies* 3: 197–212.

Punter, J. V. (1990) *Design Control in Bristol 1940–1990: The Impact of Planning on the Design of Office Development in the City Centre*, Bristol: Redcliffe.

Slater, T. R. (ed.) (1990) *The Built Form of Western Cities*, Leicester: Leicester University Press.

Steadman, P., Brown, F. and Rickaby, P. (1991) 'Studies in the morphology of the English building stock', *Environment and Planning B: Planning and Design* 18: 85–98.

Stevens, G. (1990) 'An alliance confirmed: planning literature and the social sciences', *Journal of the American Planning Institute* 56: 341–9.

Vance, J. E. (1990) *The Continuing City: Urban Morphology in Western Civilization*, Baltimore: Johns Hopkins University Press.

Vilagrasa, J. (1990) 'The fringe-belt concept in a Spanish context: the case of Lleida', in Slater, T. R. (ed.) (1990) *The Built Form of Western Cities*, Leicester: Leicester University Press.

Whitehand, J. W. R. (1988) 'Urban fringe belts: development of an idea', *Planning Perspectives* 3, 1: 47–58.

Whitehand, J. W. R. (1991) *The Making of the Urban Landscape*, Institute of British Geographers Special Publication 26, Oxford: Blackwell.

Whitehand, J. W. R. (1992) 'Recent advances in urban morphology', *Urban Studies*, 29: 617–34.

Yiftachel, O. (1989) 'Boundary change and institutional conflict in the planning of central Perth', *New Zealand Geographer* 45: 58–67.

INDEX

Note: Chapter notes and references are not indexed. References in italics refer to figures and tables.

Hampstadt, Berlin (Germany) 150
Handlin, D.P. 171
Harris, B. 310
Harvey, D.W. 217–8, 220–1, 308
Haslam, J. 44
Hastings, H. de C. 138, 141
Hayden, D. 171
Hayward Gallery, London (UK) 150
Heasly, A.E. 183
hedgerows 249, 252, 262–3
Hedon (UK) 45
Heighway, C. 55
Heikkila, E. 8
Hennell, T. 147
heritage 208; national 148
Herron, R. 150
Hervines Road, Amersham (UK) 243,
    245, 247–50, 261
highway(s) 264; authorities 233;
    dimensions and geometry of 306;
    engineers 134; matters 261;
    objections 262
high-tech corridors 8
Hill, D. 55, 57, 60
Hillingdon, London Borough of (UK)
    238, 239, 240
Hindle, B.P. 45
Hinks, R. 140
Historic Seattle Preservation and
    Development Agency 183
Historic Town Atlases 23, 44, 49
historical: expressiveness 307;
    preservation, policy on 286; record
    302; standards 294; styles 147;
    townscapes (see also historical
    urban landscapes) 6, 7, 40; urban
    landscapes (see also historical
    townscapes) 10
historicity 6, 7, 201, 307; social and
    aesthetic aspects of 201
historico-geographical theory of urban
    form 39
historiographical enquiries 136
Holdsworth, D.W. 12, 115, 120, 126,
    300
Holland Report 238
Holt, R.A. 54, 62
homeowner's association 196
Hook, Hampshire (UK) 152

Hooke, D. 60
hotel 248
Hopkins, J.S.P. 8
house(s) 149, 173, 173, 178–80, 193,
    196–7, 211, 252, 260; first-cycle
    240, 242, 254–5; forms 171–3,
    194; plans, typical 204; house
    types 175;
    apartment house 184;
    bungalow 183, 194, 187;
    condominium 191–2; detached
    172, 190, 193, 198, 241, 243,
    246, 247–9, 251–2, 254–5,
    257–8, 262–3; double house
    176, 184, 190; duplex see
    house types, double house;
    fourplex 189–90; garden
    apartment 173, 176, 189, 191,
    193, 195, 197–8, 200;
    maisonette 151, 233, 243, 252;
    mews 250; model 202; narrow
    201–2; narrow and deep 173,
    184, 193–4, 197; ranch 175,
    187, 188, 194; semi-detached
    172, 176, 187, 190, 202; senior
    212; single see house, detached;
    single-family 170, 172, 183–4,
    190, 197, 199, 201, 212;
    split-level 175, 187, 187, 188,
    190–1, 194; terraced 172, 240,
    242, 249–50; triplex 189;
    town-house 212, 240; wide and
    shallow single detached 173,
    175, 194, 197, 201; zero-lot-line
    173, 176, 190, 192, 195,
    197
housebuilder(s) 228; speculative 291
housebuilding firms 218
household(s) 235; size 227;
    upper-middle-class 222
housing 152, 171–2, 215, 227, 267;
    architect-designed 227;
    associations 238; boom 271; crisis
    238; council 242; forms, common
    172; market, slump in 250;
    revival of 252;
    middle-class 172; multi-family
    209; municipal 242; single-family
    209, 217, 254; patterns 151;